Studies in Computational Intelligence

Volume 821

Series editor

Janusz Kacprzyk, Polish Academy of Sciences, Warsaw, Poland
e-mail: kacprzyk@ibspan.waw.pl

The series "Studies in Computational Intelligence" (SCI) publishes new developments and advances in the various areas of computational intelligence—quickly and with a high quality. The intent is to cover the theory, applications, and design methods of computational intelligence, as embedded in the fields of engineering, computer science, physics and life sciences, as well as the methodologies behind them. The series contains monographs, lecture notes and edited volumes in computational intelligence spanning the areas of neural networks, connectionist systems, genetic algorithms, evolutionary computation, artificial intelligence, cellular automata, self-organizing systems, soft computing, fuzzy systems, and hybrid intelligent systems. Of particular value to both the contributors and the readership are the short publication timeframe and the world-wide distribution, which enable both wide and rapid dissemination of research output.

The books of this series are submitted to indexing to Web of Science, EI-Compendex, DBLP, SCOPUS, Google Scholar and Springerlink.

More information about this series at http://www.springer.com/series/7092

Rafał Scherer

Computer Vision Methods for Fast Image Classification and Retrieval

 Springer

Rafał Scherer
Institute of Computational Intelligence
Częstochowa University of Technology
Częstochowa, Poland

ISSN 1860-949X   ISSN 1860-9503 (electronic)
Studies in Computational Intelligence
ISBN 978-3-030-12194-5 (hardcover)   ISBN 978-3-030-12195-2 (eBook)
ISBN 978-3-030-12197-6 (softcover)
https://doi.org/10.1007/978-3-030-12195-2

Library of Congress Control Number: 2018968376

This Springer imprint is published by the registered company Springer Nature Switzerland AG
The registered company address is: Gewerbestrasse 11, 6330 Cham, Switzerland

Preface

Computer vision and image retrieval and classification are a vital set of methods used in various engineering, scientific and business applications. In order to describe an image, visual features must be detected and described. Usually, the description is in the form of vectors. The book presents methods for accelerating image retrieval and classification in large datasets. Some of the methods (Chap. 5) are designed to work directly in relational database management systems.

The book is the result of collaboration with colleagues from the Institute of Computational Intelligence at the Częstochowa University of Technology. I would like to thank my former Ph.D. students Dr. Rafał Grycuk and Dr. Patryk Najgebauer for their cooperation.

I would like to express my sincere thanks to my friend Prof. Marcin Korytkowski for his invaluable help in research and to Prof. Leszek Rutkowski, who introduced me to scientific work and supported me in a friendly manner. I am also grateful to the Institute of Computational Intelligence at the Częstochowa University of Technology for providing a scholarly environment for both teaching and research.

Finally, I am truly grateful to my wife Magda, my children Karolina and Katarzyna for their love and patience and to my mother for raising me in the way that she did.

Częstochowa, Poland
November 2018

Rafał Scherer

Contents

Chapter 1
Introduction

In recent times, one can observe the increasing development of multimedia technologies and their rising dominance in life and business. Society is becoming more eager to use new solutions as they facilitate life, primarily by simplifying contact and accelerating the exchange of experience with others, what was not encountered on such a large scale many years ago.

Computer vision solutions are being developed increasingly to oversee production processes in order to ensure their correct operation. Until now, most of them could only be properly supervised by humans. Control requires focusing and consists in constantly performing identical activities. Work monotony lowers human concentration, which is more likely to make a mistake or overlook important facts.

Healthcare, and in particular medical diagnostics, is one of the areas that provide a relatively broad spectrum of possible applications for computer vision solutions. In the past, most methods focused on processing and delivery of results in the most readable form to the doctor's diagnosis for analysis. These include medical imaging, such as computed tomography, magnetic resonance and ultrasonography, which transform signals from the device into a diagnostic readable image. Now, the diagnosis can be automatised thanks to image classification.

The most popular way to search vast collections of images and video which are generated every day in a tremendous amount is realized by keywords and meta tags or just by browsing them. Emergence of content-based image retrieval (CBIR) in the 1990s enabled automatic retrieval of images to a certain extent. Various CBIR tasks include searching for images similar to the query image or retrieving images of a certain class [11, 20, 21, 28, 29, 31, 41, 50, 51, 53] and classification [2, 6, 10, 18, 19, 22, 30, 44, 52] of the query image. Such content-based image matching remains a challenging problem of computer science. Image matching consists of two relatively difficult tasks: identifying objects in images and fast searching through large collections of identified objects. Identifying objects on images is still

© Springer Nature Switzerland AG 2020

R. Scherer, *Computer Vision Methods for Fast Image Classification and Retrieval*, Studies in Computational Intelligence 821,
https://doi.org/10.1007/978-3-030-12195-2_1

a challenge as the same objects and scenes can be viewed under different imaging conditions. There are many previous works dedicated to the problem formulated in this way. Some of them are based on color representation [15, 25, 39], textures [9, 12, 17, 46], shape [16, 23, 49] or edge detectors [37, 38, 56]. Local invariant features have gained a wide popularity [32–34, 36, 45]. The most popular local keypoint detectors and descriptors are SURF [4], SIFT [32] or ORB [42].

In content-based image retrieval and classification, we can distinguish two approaches. The first one gradually generalises information from an image. To this group, we can include methods based on machine learning such as convolutional neural networks, e.g. [8, 26] or older methods based on histograms [40, 48]. These methods try to reduce the amount of visual feature data to describe the entire image at the highest possible level. Neural networks can be trained to recognise and classify particular elements of an image, but they lose some information that is crucial to determine if the content between images is identical.

To check similarity between images we can use methods from the second popular group that is based on local interest points (keypoints), or other features that describe the local content of an image. Such methods do not generalise the content of an image and do not try to classify it. They usually generate significant amount of data, but they can find similar fragments of content between images. Thanks to this, this group of methods found multiple applications in video tracking and processing, for example, to correct content transition between frames during the camera move [14, 55]. Another popular application is a three-dimensional object reconstruction from a set of images. Some popular methods include SIFT, SURF, HOG, ORB, BRIEF, FREAK, with many modifications [3, 5, 24, 27, 43, 47, 54].

In the case of the first group of methods, work with a larger set of images is easier, because the result features are simple and in most cases can be easily stored and searched. But in the case of the second group, the main problem is a large and variable amount of data per image. To speed up the search process, we can use methods that can learn keypoint structure or descriptors [7, 13].

Developing content-based image comparison methods that simulate human visual perception is a very hard and complicated process. Image recognition is natural and very simple for humans but when we try to mimic the process we face many problems as it is very complicated, uses multiple hidden techniques developed during the evolution and we only have a rough sense of how the brain works. Most of them, e.g. human imagination, are currently unavailable for computer systems. Also huge knowledge, which humans acquire through the entire life is hard to store for machine learning systems and we excel in visual identification. Thus, image comparison algorithms try to extract and simplify this large amount of data from images to form a structurized description that is easy to compare for computers, such as human text writing [1, 35]. But image description is extracted only from the image pixel spatial distribution and is not supported by human imagination or knowledge. That caused that image description in most cases is not fully satisfactory for human users.

The book presents some methods for accelerating image retrieval and classification in large collections of images using so-called hand-crafted features.

References

1. Aghdam, M.H., Heidari, S.: Feature selection using particle swarm optimization in text categorization. J. Artif. Intell. Soft Comput. Res. **5**(4), 231–238 (2015)
2. Akusok, A., Miche, Y., Karhunen, J., Bjork, K.M., Nian, R., Lendasse, A.: Arbitrary category classification of websites based on image content. Comput. Intell. Mag. IEEE **10**(2), 30–41 (2015). https://doi.org/10.1109/MCI.2015.2405317
3. Alahi, A., Ortiz, R., Vandergheynst, P.: Freak: Fast retina keypoint. In: 2012 IEEE Conference on Computer Vision and Pattern Recognition, pp. 510–517. IEEE (2012)
4. Bay, H., Ess, A., Tuytelaars, T., Van Gool, L.: Speeded-up robust features (surf). Comput. Vis. Image Underst. **110**(3), 346–359 (2008)
5. Bay, H., Tuytelaars, T., Van Gool, L.: Surf: Speeded up robust features. In: Computer vision–ECCV 2006, pp. 404–417. Springer (2006)
6. Bazarganigilani, M.: Optimized image feature selection using pairwise classifiers. J. Artif. Intell. Soft Comput. Res. **1**(2), 147–153 (2011)
7. Calonder, M., Lepetit, V., Fua, P.: Keypoint signatures for fast learning and recognition. In: European Conference on Computer Vision, pp. 58–71. Springer (2008)
8. Chang, O., Constante, P., Gordon, A., Singana, M.: A novel deep neural network that uses space-time features for tracking and recognizing a moving object. J. Artif. Intell. Soft Comput. Res. **7**(2), 125–136 (2017)
9. Chang, T., Kuo, C.C.: Texture analysis and classification with tree-structured wavelet transform. Image Process. IEEE Trans. **2**(4), 429–441 (1993). https://doi.org/10.1109/83.242353
10. Chang, Y., Wang, Y., Chen, C., Ricanek, K.: Improved image-based automatic gender classification by feature selection. J. Artif. Intell. Soft Comput. Res. **1**(3), 241–253 (2011)
11. Daniel Carlos Guimaraes Pedronette, J.A., da S. Torres, R.: A scalable re-ranking method for content-based image retrieval. Inf. Sci. **265**(0), 91–104 (2014). https://doi.org/10.1016/j.ins.2013.12.030
12. Francos, J., Meiri, A., Porat, B.: A unified texture model based on a 2-d wold-like decomposition. Signal Process. IEEE Trans. **41**(8), 2665–2678 (1993). https://doi.org/10.1109/78.229897
13. Grabner, M., Grabner, H., Bischof, H.: Learning features for tracking. In: IEEE Conference on Computer Vision and Pattern Recognition, 2007 CVPR 2007, pp. 1–8. IEEE (2007)
14. Hare, S., Saffari, A., Torr, P.H.: Efficient online structured output learning for keypoint-based object tracking. In: 2012 IEEE Conference on Computer Vision and Pattern Recognition (CVPR), pp. 1894–1901. IEEE (2012)
15. Huang, J., Kumar, S., Mitra, M., Zhu, W.J., Zabih, R.: Image indexing using color correlograms. In: Proceedings of 1997 IEEE Computer Society Conference on Computer Vision and Pattern Recognition, 1997, pp. 762–768 (1997). https://doi.org/10.1109/CVPR.1997.609412
16. Jagadish, H.V.: A retrieval technique for similar shapes. SIGMOD Rec. **20**(2), 208–217 (1991)
17. Jain, A.K., Farrokhnia, F.: Unsupervised texture segmentation using gabor filters. Pattern Recogn. **24**(12), 1167–1186 (1991)
18. Jégou, H., Douze, M., Schmid, C., Pérez, P.: Aggregating local descriptors into a compact image representation. In: 2010 IEEE Conference on Computer Vision and Pattern Recognition (CVPR), pp. 3304–3311. IEEE (2010)
19. Jégou, H., Perronnin, F., Douze, M., Sanchez, J., Perez, P., Schmid, C.: Aggregating local image descriptors into compact codes. Pattern Anal. Mach. Intell. IEEE Trans. **34**(9), 1704–1716 (2012)
20. Kanimozhi, T., Latha, K.: An integrated approach to region based image retrieval using firefly algorithm and support vector machine. Neurocomputing **151, Part 3**(0), 1099–1111 (2015)
21. Karakasis, E., Amanatiadis, A., Gasteratos, A., Chatzichristofis, S.: Image moment invariants as local features for content based image retrieval using the bag-of-visual-words model. Pattern Recogn. Lett. **55**, 22–27 (2015)

22. Karimi, B., Krzyzak, A.: A novel approach for automatic detection and classification of suspicious lesions in breast ultrasound images. J. Artif. Intell. Soft Comput. Res. **3**(4), 265–276 (2013)
23. Kauppinen, H., Seppanen, T., Pietikainen, M.: An experimental comparison of autoregressive and fourier-based descriptors in 2d shape classification. Pattern Anal. Mach. Intell. IEEE Trans. **17**(2), 201–207 (1995). https://doi.org/10.1109/34.368168
24. Ke, Y., Sukthankar, R.: Pca-sift: A more distinctive representation for local image descriptors. In: Proceedings of the 2004 IEEE Computer Society Conference on Computer Vision and Pattern Recognition, 2004. CVPR 2004. , vol. 2, pp. II–II. IEEE (2004)
25. Kiranyaz, S., Birinci, M., Gabbouj, M.: Perceptual color descriptor based on spatial distribution: a top-down approach. Image Vision Comput. **28**(8), 1309–1326 (2010)
26. Krizhevsky, A., Sutskever, I., Hinton, G.E.: Imagenet classification with deep convolutional neural networks. Adv. Neural Inf. Process. Syst., 1097–1105 (2012)
27. Leutenegger, S., Chli, M., Siegwart, R.Y.: Brisk: Binary robust invariant scalable keypoints. In: 2011 IEEE International Conference on Computer Vision (ICCV), pp. 2548–2555. IEEE (2011)
28. Lin, C.H., Chen, H.Y., Wu, Y.S.: Study of image retrieval and classification based on adaptive features using genetic algorithm feature selection. Expert Syst. Appl. **41**(15), 6611–6621 (2014)
29. Liu, G.H., Yang, J.Y.: Content-based image retrieval using color difference histogram. Pattern Recogn. **46**(1), 188–198 (2013)
30. Liu, L., Shao, L., Li, X.: Evolutionary compact embedding for large-scale image classification. Inf. Sci. **316**, 567–581 (2015)
31. Liu, S., Bai, X.: Discriminative features for image classification and retrieval. Pattern Recogn. Lett. **33**(6), 744–751 (2012)
32. Lowe, D.G.: Distinctive image features from scale-invariant keypoints. Int. J. Comput. Vis. **60**(2), 91–110 (2004)
33. Matas, J., Chum, O., Urban, M., Pajdla, T.: Robust wide-baseline stereo from maximally stable extremal regions. Image Vis. Comput. **22**(10), 761–767 (2004). British Machine Vision Computing 2002
34. Mikolajczyk, K., Schmid, C.: Scale and affine invariant interest point detectors. Int. J. Comput. Vis. **60**(1), 63–86 (2004)
35. Murata, M., Ito, S., Tokuhisa, M., Ma, Q.: Order estimation of japanese paragraphs by supervised machine learning and various textual features. J. Artif. Intell. Soft Comput. Res. **5**(4), 247–255 (2015)
36. Nister, D., Stewenius, H.: Scalable recognition with a vocabulary tree. In: Proceedings of the 2006 IEEE Computer Society Conference on Computer Vision and Pattern Recognition - Volume 2, CVPR '06, pp. 2161–2168. IEEE Computer Society, Washington, DC, USA (2006)
37. Ogiela, M.R., Tadeusiewicz, R.: Syntactic reasoning and pattern recognition for analysis of coronary artery images. Artif. Intell. Med. **26**(1), 145–159 (2002)
38. Ogiela, M.R., Tadeusiewicz, R.: Nonlinear processing and semantic content analysis in medical imaging-a cognitive approach. Instrum. Measurement IEEE Trans. **54**(6), 2149–2155 (2005)
39. Pass, G., Zabih, R.: Histogram refinement for content-based image retrieval. In: Proceedings 3rd IEEE Workshop on Applications of Computer Vision, 1996. WACV 1996, pp. 96–102 (1996). https://doi.org/10.1109/ACV.1996.572008
40. Pass, G., Zabih, R., Miller, J.: Comparing images using color coherence vectors. In: Proceedings of the fourth ACM international conference on Multimedia, pp. 65–73. ACM (1997)
41. Rashedi, E., Nezamabadi-pour, H., Saryazdi, S.: A simultaneous feature adaptation and feature selection method for content-based image retrieval systems. Knowl. Based Syst. **39**, 85–94 (2013)
42. Rublee, E., Rabaud, V., Konolige, K., Bradski, G.: Orb: An efficient alternative to sift or surf. In: 2011 IEEE International Conference on Computer Vision (ICCV), pp. 2564–2571 (2011). https://doi.org/10.1109/ICCV.2011.6126544
43. Rublee, E., Rabaud, V., Konolige, K., Bradski, G.: Orb: an efficient alternative to sift or surf. In: 2011 IEEE International Conference on Computer Vision (ICCV), pp. 2564–2571. IEEE (2011)

44. Shrivastava, N., Tyagi, V.: Content based image retrieval based on relative locations of multiple regions of interest using selective regions matching. Inf. Sci. **259**, 212–224 (2014). https://doi.org/10.1016/j.ins.2013.08.043
45. Sivic, J., Zisserman, A.: Video google: a text retrieval approach to object matching in videos. In: Proceedings of Ninth IEEE International Conference on Computer Vision, 2003, pp. 1470–1477 vol. 2 (2003)
46. Śmietański, J., Tadeusiewicz, R., Łuczyńska, E.: Texture analysis in perfusion images of prostate cancera case study. Int. J. Appl. Math. Comput. Sci. **20**(1), 149–156 (2010)
47. Sünderhauf, N., Protzel, P.: Brief-gist-closing the loop by simple means. In: 2011 IEEE/RSJ International Conference on Intelligent Robots and Systems (IROS), pp. 1234–1241. IEEE (2011)
48. Tsai, G.: Histogram of oriented gradients. Univ. Mich. **1**(1), 1–17 (2010)
49. Veltkamp, R.C., Hagedoorn, M.: State of the art in shape matching. In: Lew, M.S. (ed.) Principles of Visual Information Retrieval, pp. 87–119. Springer, London, UK, UK (2001)
50. Wang, X.Y., Yang, H.Y., Li, Y.W., Li, W.Y., Chen, J.W.: A new svm-based active feedback scheme for image retrieval. Eng. Appl. Artif. Intell. **37**, 43–53 (2015)
51. Wu, J., Shen, H., Li, Y.D., Xiao, Z.B., Lu, M.Y., Wang, C.L.: Learning a hybrid similarity measure for image retrieval. Pattern Recogn. **46**(11), 2927–2939 (2013)
52. Yang, J., Yu, K., Gong, Y., Huang, T.: Linear spatial pyramid matching using sparse coding for image classification. In: IEEE Conference on Computer Vision and Pattern Recognition, 2009. CVPR 2009, pp. 1794–1801 (2009). https://doi.org/10.1109/CVPR.2009.5206757
53. Yu, J., Qin, Z., Wan, T., Zhang, X.: Feature integration analysis of bag-of-features model for image retrieval. Neurocomputing **120**(0), 355–364 (2013). Image Feature Detection and Description
54. Žemgulys, J., Raudonis, V., Maskeliūnas, R., Damaševičius, R.: Recognition of basketball referee signals from videos using histogram of oriented gradients (hog) and support vector machine (svm). Procedi. Comput. Sci. **130**, 953–960 (2018)
55. Zhao, W.L., Ngo, C.W.: Scale-rotation invariant pattern entropy for keypoint-based near-duplicate detection. IEEE Trans. Image Process. **18**(2), 412–423 (2009)
56. Zitnick, C., Dollar, P.: Edge boxes: Locating object proposals from edges. In: Fleet, D., Pajdla, T., Schiele, B., Tuytelaars, T., (eds.) Computer Vision ECCV 2014, Lecture Notes in Computer Science, vol. 8693, pp. 391–405. Springer International Publishing (2014)

Chapter 2
Feature Detection

Computer vision relies on image features describing points, edges, objects or colour. The book concerns solely so-called hand-made features contrary to learned features which exist in deep learning methods. Image features can be generally divided into global and local methods.

Global methods extract features from the entire image without dividing into more and less significant areas. To this group, we can include histogram-based algorithms such as histogram of oriented gradients (HOG) or colour coherence vector (CCV) [12, 47]. In most cases, they generate a constant amount of description data which is easier to compare and store, on the other hand, image comparison by histogram-based algorithms gives only a vague similarity for a user.

Local feature-based methods try at first to find significant characteristic areas of an image based on Laplacian of Gaussian (LoG) or Difference of Gaussian (DoG) algorithms [25, 64]. And then they generate a description of their neighbourhood. These methods are more accurate, on the other hand, can generate a lot of description data and that amount varies per image. Local feature methods based on keypoints are efficient in similarity detection between images but less in content recognition. Commonly used methods of this kind are SIFT, SURF, ORB, BRIEF, FAST [4, 6, 46, 48, 49].

2.1 Local Features

2.1.1 Scale-Invariant Feature Transform (SIFT)

SIFT (Scale-Invariant Feature Transform) is an algorithm used to detect and describe local features of an image. It was presented for the first time in [37] and is now

© Springer Nature Switzerland AG 2020

R. Scherer, *Computer Vision Methods for Fast Image Classification and Retrieval*, Studies in Computational Intelligence 821,

https://doi.org/10.1007/978-3-030-12195-2_2

patented by the University of British Columbia. For each keypoint, which describes the local image feature, we generate a feature vector, that can be used for further processing. The algorithm is immune to changing scale, rotation and light change. SIFT consists of the four main steps [36]

1. Scale-space extreme detection—Extraction of potential keypoints by scanning the entire image,

   Constructing scale-space,
   Laplacian approximation by Gaussian blur,

2. Keypoint localization—Selection of stable keypoints (resistant to change of scale and rotation),

   Removing not important keypoints (noise).

3. Orientation assignment—Finding keypoint orientation resistant to the image transformation,
4. Keypoint descriptor—Generating vectors describing keypoints.

During the process of creating scale-space, the image is rescaled (creating octaves) in order to detect the most important and resistant features. After this step, a scale-space pyramid is obtained. This pyramid consists of octaves, sorted from the largest to the smallest octave.

In the next stage, Gaussian blur is applied. This step is performed by the following Gaussian operator [37]

$$L(x, y, \sigma) = G(x, y, \sigma) * I(x, y), \tag{2.1}$$

where L is the output image, G represents the Gaussian operator, I is the input image. In the next step, the Laplacian is calculated, in order to detect edges. It should be done by calculation of the second derivative, but this operation is computationally expensive. In order to overcome this nuisance, the Difference of Gaussians (DOG) is performed. The next stage of the SIFT algorithm is the localisation of keypoints. This step consists of two important steps

- localisation of local extrema on DoG images,
- setting the extrema position.

The localisation of local extrema is based on comparing the pixels with their neighbours. On a discrete image, the brightest pixel not always has the same position as local extrema. Thus, this issue was solved by using Taylor's theorem

$$D(x) = D + \frac{\partial D^T}{\partial x}x + \frac{1}{2}x^T\frac{\partial^2 D}{\partial x^2}x. \tag{2.2}$$

Each keypoint contains two features: strength and orientation in which keypoint is directed. It is calculated by gradients of its neighbours. Each resultant SIFT keypoint descriptor consists of two vectors. The first one contains the point position (x, y), scale (detected scale), response (response of the detected feature, strength), orientation (orientation measured anti-clockwise from +ve x-axis), the Laplacian (sign of the Laplacian for fast matching purposes). The second one contains the descriptor of 128 length.

2.1.2 Speed-Up Robust Features (SURF)

SURF (Speeded-Up Robust Features) is a method which allows to detect and describe local features of an image. SURF is an improved version of SIFT in terms of speed (see Sect. 2.1.1) [37]. It was presented for the first time in [4], nowadays is widely used in various systems e.g. image recognition, object description [24], segmentation [23], image analysis [24] and image retrieval [22, 57], object tracking [21], and many others. SURF is similar algorithm to SIFT. The Integral Images are used instead of DOG (Difference of Gaussian), which allows it to work much faster than SIFT. An important advantage of SURF is that it generates less data than SIFT (SURF has a shorter descriptor of 64 length), which speeds up further processing. The algorithm has also a parallel version [53, 56] thus, it generates the results much faster. SURF generates image keypoints (interesting points) which allows to extract and match local features, e.g. in pairs of images. For each keypoint, which indicates a local image feature, a feature vector is generated. It describes the keypoint surrounding and allows to determine its orientation. These vectors are often used for further processing in many computer vision methods. SURF consists of four main steps:

- Computing Integral Images,
- Fast Hessian Detector,

  – The Hessian,
  – Constructing the Scale-Space,
  – Accurate Interest Point Localization,

- Interest Point Descriptor,

  – Orientation Assignment,
  – Descriptor Components,

- Generating vectors describing the keypoint.

In the first step, the integral images are calculated. It allows to increase the efficiency. This method is very simple and it is used for calculation of pixels sum in the given rectangle area. This process can be described by the following formula [15]

$$I_{\Sigma}(x, y) = \sum_{i=0}^{i<x} \sum_{j=0}^{j<y} I(x, y), \tag{2.3}$$

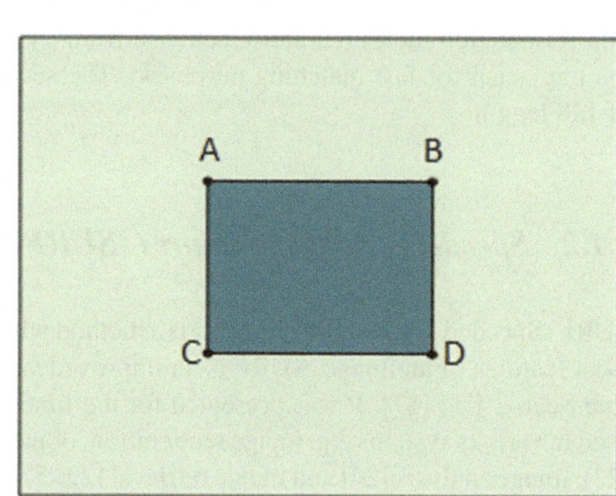

Fig. 2.1 Calculating area using integral images

where I is a processed image and $I_\Sigma(x, y)$ is a sum of pixels in given area. The usage of integral images in order to calculating area is reduced to four operation. Let us consider a rectangle described by vertices A, B, C, D. The example is presented in Fig. 2.1. The sum of pixel intensities is calculated by formula [15]

$$\sum = A + D - (C + B). \tag{2.4}$$

In the next step, the Hessian matrix determinant is calculated. The Hessian matrix is presented below [15]

$$H(f(x, y)) = \begin{bmatrix} \frac{\partial^2 f}{\partial x^2} & \frac{\partial^2 f}{\partial x \partial y} \\ \frac{\partial^2 f}{\partial x \partial y} & \frac{\partial^2 f}{\partial y^2} \end{bmatrix}. \tag{2.5}$$

The determinant of this matrix can be calculated by [15]

$$det(H) = \frac{\partial^2 f}{\partial x^2} \frac{\partial^2 f}{\partial y^2} - \left(\frac{\partial^2 f}{\partial x \partial y} \right). \tag{2.6}$$

The calculation of the local maximum by the Hessian matrix determinant depends from the sign of this determinant. If its value is greater or equal 0, this area is determined as the local maximum. In the next step, the scale-space is constructed. This step is used in order to make the keypoint immune to changing scale and rotation. In the interest point localization stage the $minHessian$ parameter is needed. To determine the threshold value. The localization is calculated by comparing the Hessian determinant with its neighbours.

The process of creating the keypoint descriptors is performed by using Haar wavelets (see Fig. 2.2) which describe the keypoints gradients. In order to compute the descriptor orientation, the algorithm searches the largest Haar wavelets sum in $\pi/3$ (60) window and step $+-15$ (see Fig. 2.3).

Fig. 2.2 Haar wavelets. These filters calculate responses for x (left) and y (right) directions

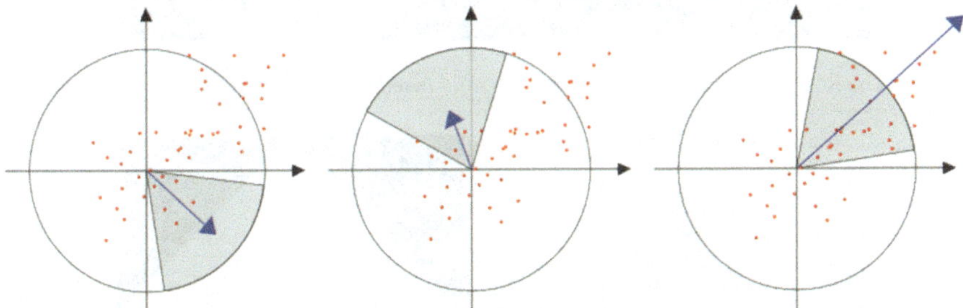

Fig. 2.3 Descriptor orientation assignment. The window of $\pi/3$ size moves around the origin and determines the sum of largest wavelets sum, which allows obtaining the longest vector [15]

Fig. 2.4 Visual representation of the SURF descriptor. The descriptor contains 16 subregions and each of them consists of 5×5 calculated Haar wavelets [15]

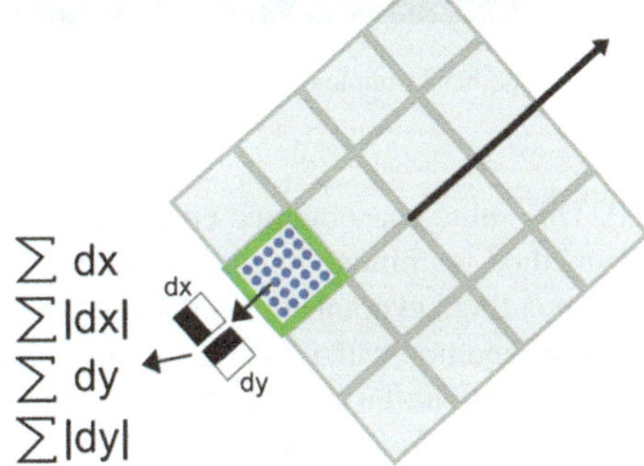

The descriptor consists of 4×4 wavelet group matrix. Each of them is composed of the calculated Haar wavelets and is divided into 5×5 elements (Fig. 2.4).

The output SURF keypoint consists of two vectors. The first one contains: point position (x, y), scale (detected scale), response (response of the detected feature, strength), orientation (orientation measured anti-clockwise from +ve x-axis), the Laplacian (the sign of the Laplacian for fast matching purposes). The second one describes the intensity distribution of the pixels within the neighbourhood of the interest point (64 values). In order to generate keypoints, SURF requires one input parameter—$minHessian$. During the experiments, this variable was usually set to

Fig. 2.5 The SURF algorithm example with keypoint detection and matching

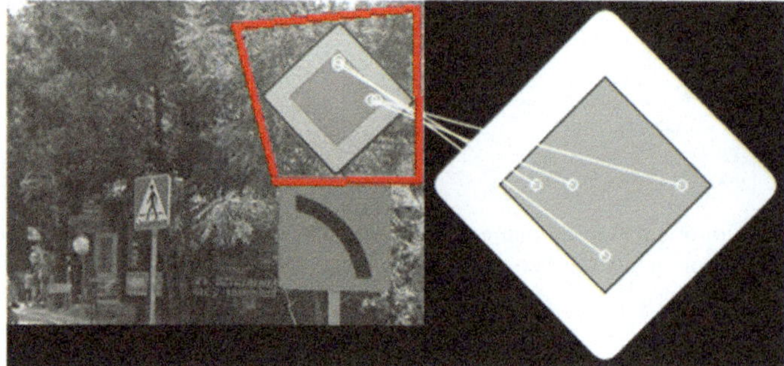

Fig. 2.6 Another example of keypoint matching by the SURF algorithm

400. This value was obtained empirically, and it was suitable in many experiments. The method is immune to change of scale and rotation, which allows for matching corresponding keypoints in similar images [27, 43]. Figure 2.5 shows an example of the SURF method with two images with similar objects. The lines between these two images represent the corresponding keypoints found on both images. The rectangle on the observed image pinpoints the object location (Fig. 2.6).

2.2 Edge Detection

Edge detection [7] is one of the most commonly used stages in image processing [30]. This step is extremely relevant in the image description process. It allows detecting the object shape, which is crucial in further steps of image classification or retrieval. Usually, edges are detected between two different regions in the image (gradient change). They occur when the image intensity or its first derivative changes. The visual representation of derivatives is presented in Fig. 2.7.

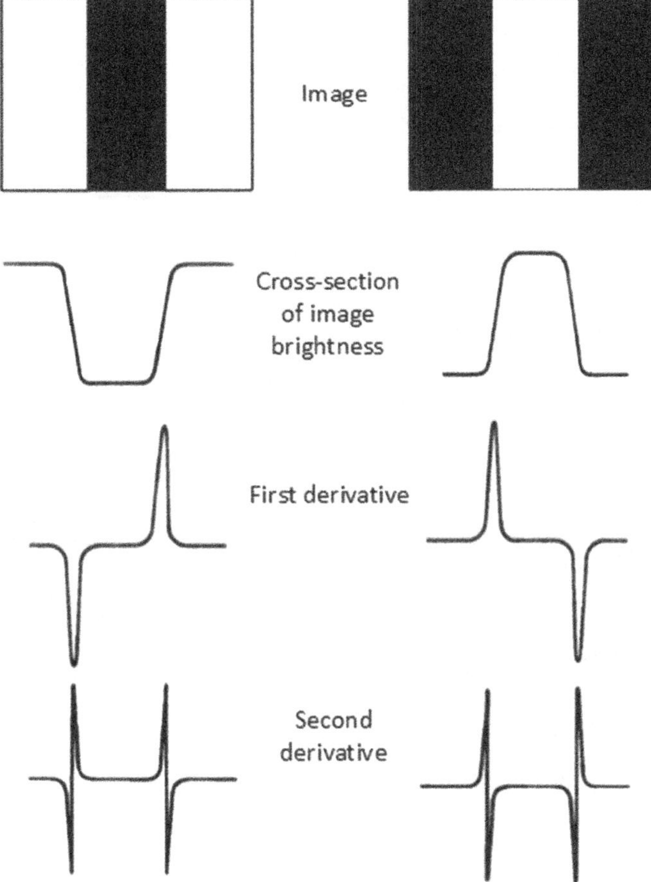

Fig. 2.7 Examples of image derivatives

There are four types of edges (see Fig. 2.8) [30]

- Ramp—edge values increase gradually. This edge profile is commonly used in photos. It can be also found after image blur process,
- Step—intensity value changes abruptly, from one side to another. This is the most desirable edge profile. They are often observed in computer images,
- Line—this edge profile is composed of two closely positioned edges. They are detected by the first derivative filters and they are immune to most blur techniques,
- Roof—this profile is frequently observed in lines on the photographic images.

Essentially, the edge detection process is based on detecting significant local changes in the image. Let us take for consideration one dimension, the step edge is related to a local peak of the first derivative. In such case, the gradient is defined as the measure of change of a function. It can also be considered as an array of continuous function

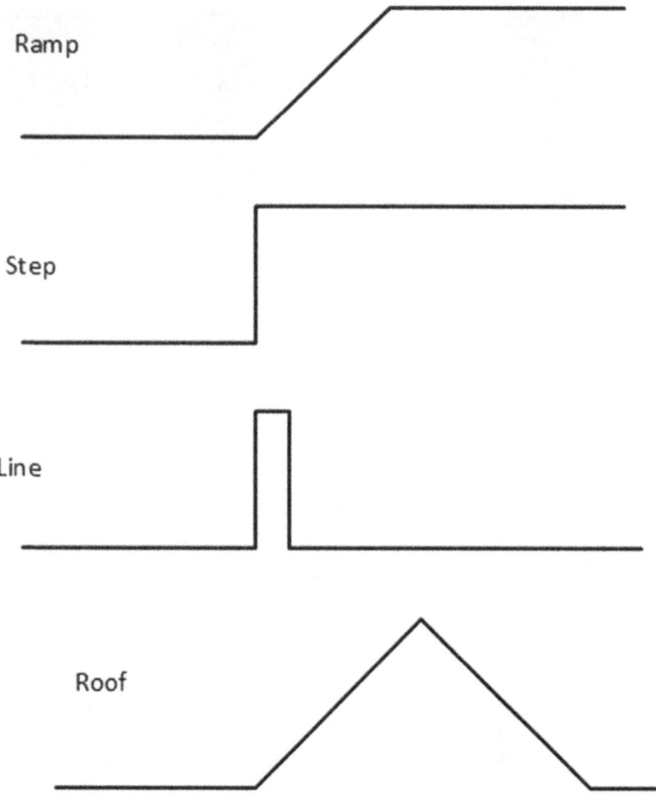

Fig. 2.8 One-dimensional edge profiles

samples of image intensity. Thus, the gradient is equivalent of the first derivative and it is described by [30]

$$G[f(x, y)] = \begin{bmatrix} G_x \\ G_y \end{bmatrix} = \begin{bmatrix} \frac{\partial f}{\partial x} \\ \frac{\partial f}{\partial y} \end{bmatrix}, \tag{2.7}$$

where G_x and G_y is expressed by filters. In order to approximate the gradient in both directions (x, y) the following filters can be used [2, 52]

- the Sobel operator

$$Gx = \begin{bmatrix} -1 & 0 & 1 \\ -2 & 0 & 2 \\ -1 & 0 & 1 \end{bmatrix}, \tag{2.8}$$

$$Gy = \begin{bmatrix} 1 & 2 & 1 \\ 0 & 0 & 0 \\ -1 & -2 & -1 \end{bmatrix}, \tag{2.9}$$

- the Prewit filter

$$Gx = \begin{bmatrix} -1 & 0 & 1 \\ -1 & 0 & 1 \\ -1 & 0 & 1 \end{bmatrix}, \tag{2.10}$$

$$Gy = \begin{bmatrix} 1 & 1 & 1 \\ 0 & 0 & 0 \\ -1 & -1 & -1 \end{bmatrix}, \tag{2.11}$$

- the Robert's Cross operator

$$Gx = \begin{bmatrix} 0 & 1 \\ -1 & 0 \end{bmatrix}, \tag{2.12}$$

$$Gy = \begin{bmatrix} 1 & 0 \\ 0 & -1 \end{bmatrix}, \tag{2.13}$$

- the Scharr's operator

$$Gx = \begin{bmatrix} -3 & 0 & 3 \\ -10 & 0 & 10 \\ -3 & 0 & 3 \end{bmatrix}, \tag{2.14}$$

$$Gy = \begin{bmatrix} 3 & 10 & 3 \\ 0 & 0 & 0 \\ -3 & -10 & -3 \end{bmatrix}. \tag{2.15}$$

In order to determine the edge strengths (the gradient magnitude), we can use the Euclidean (2.16) distance measure [38]

$$|G| = \sqrt{G_x^2 + G_y^2}. \tag{2.16}$$

It is common practice to approximate this magnitude by the absolute values

$$|G| \approx |G_x| + |G_y|, \tag{2.17}$$

or

$$|G| \approx max(|G_x|, |G_y|), \tag{2.18}$$

where G_x is horizontal direction gradient and G_y is vertical direction gradient. The edge direction (angle) is determined by the following formula [7, 38]

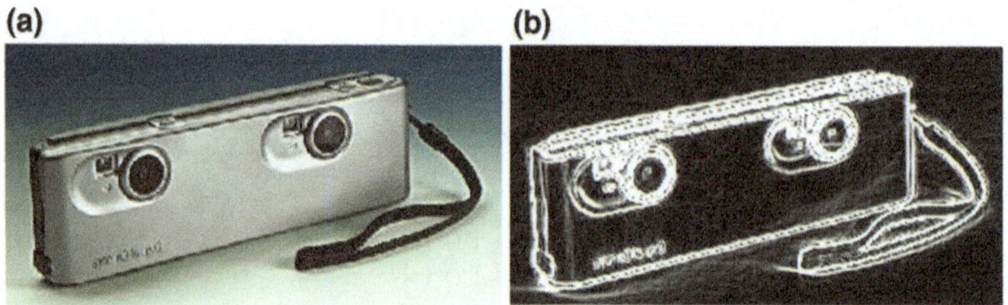

Fig. 2.9 The edge detection by the Sobel filters. Figure 2.9a—input image, Fig. 2.9b—edge detection (input image taken from the PASCAL VOC dataset [16])

$$\Theta = \arctan\left(\frac{G_x}{G_y}\right), \tag{2.19}$$

where Θ is measured with respect to the x axis. It should be noted, that the gradient magnitude is independent of the direction of the edge. An example of edge detection by the Sobel filters is presented in Fig. 2.9.

2.2.1 Canny Edge Detection

Canny edge detection is one of the most commonly used image processing methods for detecting edges [3, 59], proposed by John F. Canny in 1986. It takes as input a gray scale image, and produces an image showing the positions of tracked intensity discontinuities. The algorithm is composed of 5 separate steps:

1. Noise reduction. The image input is smoothed by applying an appropriate Gaussian filter.
2. Finding the intensity gradient of the image. During this step, the edges are marked where gradients of the image have large magnitudes.
3. Non-maxima suppression. If the gradient magnitude at a pixel is larger than those at its two neighbours in the gradient direction, mark the pixel as an edge. Otherwise, mark the pixel as the background.
4. Edge tracking by hysteresis. The final edges are determined by suppressing all edges that are not connected to the genuine edges.

The result of the Canny operator is determined by the following parameters:

- Width—of the Gaussian filter used in the first stage directly affects the results of the Canny algorithm,
- Threshold (step)—used during edge tracking by hysteresis. It is difficult to provide a generic threshold that works well on all images.

The Canny detector finds edges where the pixel intensity changes (image gradient). Before edge detection, not important edges need to be removed. Thus, the Gaus-

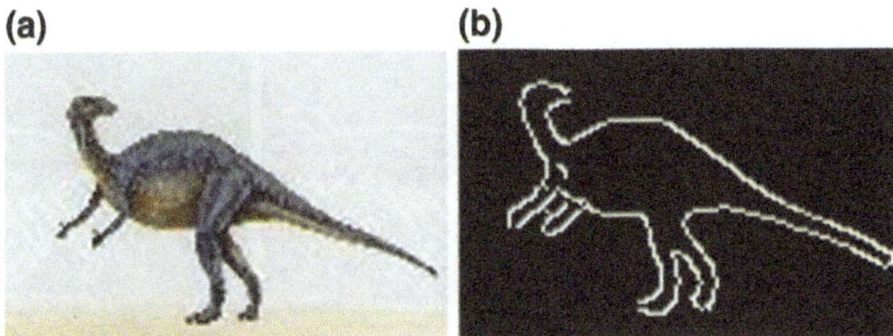

Fig. 2.10 Figure 2.10a shows an input image and Fig. 2.10b represents the edge detected image. As shown, the edges were detected correctly, because the image gradient is low (Fig. 2.10a was taken from the Corel image dataset [55])

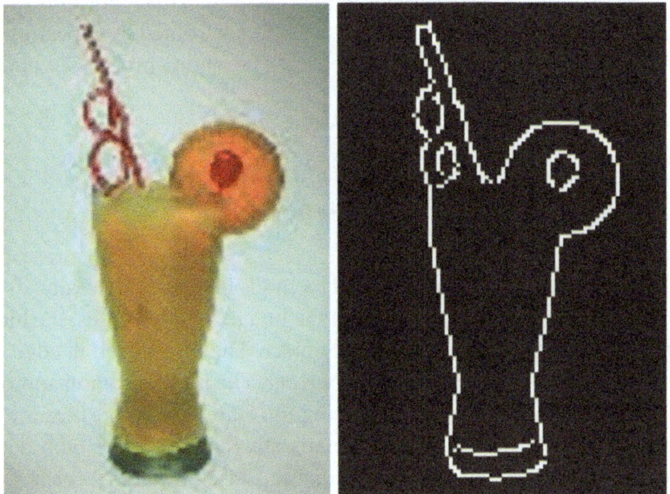

Fig. 2.11 Another example of edge detection on an image taken from the Corel dataset [55]

sian smoothing is applied. This process is performed by the Gaussian function for calculating transformation for each image pixel [38]

$$G(x, y) = \frac{1}{\sqrt{2\pi\sigma^2}} e^{-\frac{x^2+y^2}{2\sigma^2}}, \tag{2.20}$$

where x is the distance from pixel position in the horizontal axis, y is the distance from pixel position in the vertical axis, σ is the standard deviation of Gaussian distribution (Fig. 2.11).

In some cases, simple edge detection does not provide satisfying results. The edges are not always complete, and they do not describe the entire object. In order

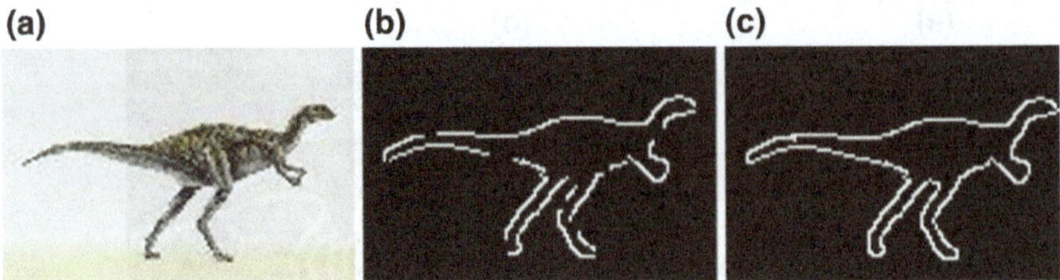

Fig. 2.12 The edge linking process. Figure 2.12a—input image, Fig. 2.12b—edge detection, Fig. 2.12c—edge linking

to eliminate this nuisance, edge linking method is used. This algorithm simply connects the edges in close proximity. This process is performed by repeating the edge detection, with a lower step value parameter. The entire process is presented in Fig. 2.12.

2.3 Blob Detection and Blob Extraction

Blob detection is one of the commonly used methods in image processing. It allows to detect and extract a list of blobs (objects) in the image. In most cases, the blob distinguishes from the background and the other object. The filters used in edge detection cannot be used due to the large size of the objects. The most common approach used as preprocessing in blob detection is applying the Gaussian blur. Thus, the obtained image is simplified, and not important details are removed. The blob detection is widely used in various areas such as medicine [44], military, forensics [63] and many others. Unfortunately, obtaining homogeneous objects from an image as a list of pixels is extremely complicated. Especially when we deal with a heterogeneous background, i.e. the objects containing multicoloured background. There are many methods for extracting objects (blobs) from images [13, 22, 24]. These algorithms are described by Andrew Kirillov [34]. The author distinguished four types of the filters: Convex full, Left/Right Edges, Top/Bottom Edges, Quadrilateral. Figure 2.13 presents these blob detection methods. Figure 2.13a illustrates Convex full filter. As can be seen, round edges of the objects are not detected correctly. Much better results are obtained by the Top/Bottom Edges filter (Fig. 2.13C). Object edges are detected mostly correctly, with individual exceptions. The Left/Right Edges method behaves similarly (Fig. 2.13b). The last method has a problem with the detection of vertices inside figures, e.g. star-shaped objects (Fig. 2.13d).

Blob extraction can be described by the following steps (Quadrilateral):

1. Locate each separate object in the input image,
2. Find object edge pixels (methods: Top/Bottom, Left/Right),
3. Detect four corners of the quadrilateral,

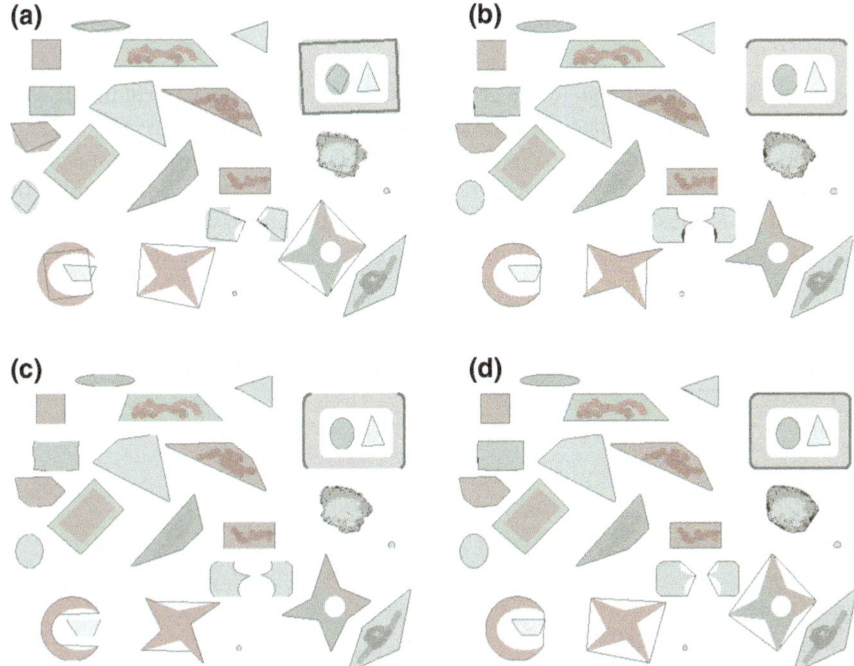

Fig. 2.13 Comparison of methods for blob detection used in the AForge.NET library [34]

4. Set $distortion Limit$. This value determines how different the object can be. Set 0 to detect perfect quadrilateral,
5. Check how well the analyzed shape fits into the quadrilateral with the assumed parameter (see Fig. 2.13d).
6. Check each mean distance between given edge pixels and the edge of assumed quadrilateral,
7. If the mean distance is not greater than $distortion Limit$, then we can assume that the shape is quadrilateral.

2.4   Clustering Algorithms

Data clustering is widely used in data mining and its applications. Clustering is a process of organising data sets into homogeneous groups. This organisation process is based on the given features of similar objects. There were many clustering algorithms proposed so far.

2.4.1 K-means Clustering Algorithm

K-means algorithm belongs to the group of heuristic clustering algorithms. It was firstly proposed by MacQueen in 1967 [40] and improved by Hartigan in 1979 [28]. The purpose of this method is to divide N points in D dimensional space into K clusters. That process of partitioning requires only one input parameter K the number of clusters [19]. K-means is commonly used for data clustering in various applications. It is relatively fast and easy in implementation. The following pseudo-code describes its functioning [32, 58]:

> **INPUT:** $X = \{x_1, x_2, ..., x_n\} \in R^n$ (data to cluster), k (number of clusters)
> **OUTPUT:** $CEN = \{cen_1, cen_2, ..., cen_n\} \in R^n$ (cluster centers),
> $L = \{l_1, l_2, ..., l_n\}$
> **for** $i = 1; i <= n; i + +$ **do**
> | $CEN[i] = rand()$;
> **end**
> **foreach** $x_i \in X$ **do**
> | $l_i := ArgMinDist(x_i, cen_j), j \in \{1..k\}$
> **end**
> $ch := false$;
> **while** $ch = false$ **do**
>   **foreach** $cen_i \in CEN$ **do**
> | $UpdateCluster(cen_i)$;
>   **end**
>   **foreach** $x_i \in X$ **do**
> $minDistance := ArgMinDist(x_i, cen_j), j \in \{1..k\}$
>   **if** $minDistance l_i$ **then**
> | $l_i := minDistance$
> | $ch := true$
>   **end**
>   **end**
> **end**

Algorithm 1: K-means algorithm.

Figure 2.14 shows the effect of colour clustering, where the left image is the input and the right one is the output. As can be seen, the number of colours has been reduced and only important features remain.

2.4.2 Mean Shift Clustering Algorithm

Mean shift is a clustering algorithm which does not require setting the number of output clusters or their shapes. The number of parameters of the algorithm is in fact, limited to the radius [11]. The basic version of the algorithm was presented in the two-dimensional Euclidean space. The task of the algorithm is to compute the

Fig. 2.14 Image colour clustering by the k-means algorithm

Euclidean distance between each point and all other points. Mean shift determines the points in d-dimensional space as a probability density function, where the denser regions correspond to local maxima. For each point in this space, there is performed a procedure of gradient increase until coverage. Points assigned to one agent of group (stationary point) are considered to be a part of the group and form a single cluster (group) [17]. Given n points $x_i \in R^d$, multivariate kernel density function $K(x)$ is expressed using the following equation [10, 11, 17]

$$\widehat{f_k} = \frac{1}{nh^d} \sum_{i=1}^{n} \left(\frac{x - x_i}{h} \right), \tag{2.21}$$

where h is a radius of the kernel function. The kernel function is defined as follows [10, 17] $K(x) = c_k k(\| x \|^2)$, where c_k is a normalization constant. If estimator density gradient is given it is possible to make the following calculations [14]

$$\nabla \widehat{f}(x) = \frac{2c_{k,d}}{nh^{d+2}} \underbrace{\left[\sum_{i=1}^{n} g \left(\| \frac{x - x_i}{h} \|^2 \right) \right]}_{term1} \underbrace{\left[\frac{\sum_{i=1}^{n} x_i g \left(\| \frac{x-x_i}{h} \|^2 \right)}{\sum_{i=1}^{n} g \left(\| \frac{x-x_i}{h} \|^2 \right)} - x \right]}_{term2}, \tag{2.22}$$

where $g(x) = -k'(x)$ is derivative of selected kernel of function. First term ($term1$) of formula 2.22 allows to define the density and second term ($term2$) is named as mean shift vector $m(x)$. Points in the direction of maximum gain and proportional to the density gradient can be determined at the point x obtained with the kernel function K.

The algorithm can be represented in the following steps [5, 14, 39]:

1. Determine mean shift vector, expressed by the formula $m(x_t^i)$,
2. Map the estimated density of window: $x_{t+1}^i = x_t^i + m(x_t^i)$,
3. Repeat the first and the second step until: $\nabla f(x_i) = 0$.

Then the group measures are calculated, and they are assigned to the selected group. This process is presented in Fig. 2.15. Figure 2.15a presents an example of the points arranged in two-dimensional space. Figure 2.15b presents the results of the algorithm and the detected six different classes.

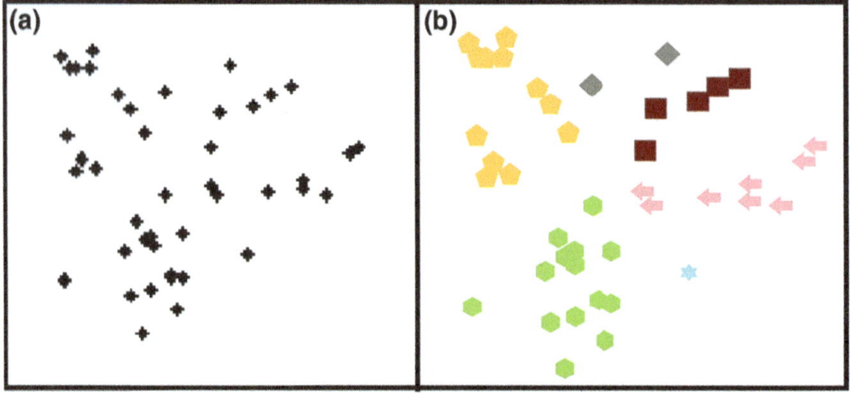

Fig. 2.15 An example of mean shift clustering. In the image A there are points before grouping and in B points grouped by the mean shift algorithm

2.5 Segmentation

Segmentation is the process of dividing the input image on the homogeneous objects having similar properties. Segmentation is essentially a division of an image into meaningful structures. It is often used in image analysis. Segmentation is one of the most complicated image processing fields, as it is challenging to obtain uniform, homogeneous objects from images containing background or overlapping areas. Many segmentation methods were proposed in the past decades, but none of them is universal for all types of images [8]. In the literature, we can also find many methods of image segmentation, which can be divided into the following groups [41]:

- Neighborhood-based segmentation—Methods that operate on image areas— homogeneity of pixel neighbourhoods,
- Threshold-based segmentation—these techniques are based on histograms and slicing in order to segment the image. It can be applied to the image directly, or some pre and post-processing can be used [1],
- Edge-based segmentation— Methods based on edge detection, consisting in determining the boundaries and contour of objects and filling them [51],
- Region-based segmentation—this technique takes the opposite approach than edge detection. The algorithm starts in the middle of the object and "grows" outward until it meets the object edges [20],
- Clustering-based segmentation—these methods use clustering algorithms, which attempt to group similar patterns (colour, texture) [61]. The two widely used clustering algorithms (k-means and mean shift) are described in Sects. 2.4.1 and 2.4.2. An example image segmented by clustering is presented in Fig. 2.14,
- Matching-based segmentation—it is based on pattern matching to a specified object [18, 50].

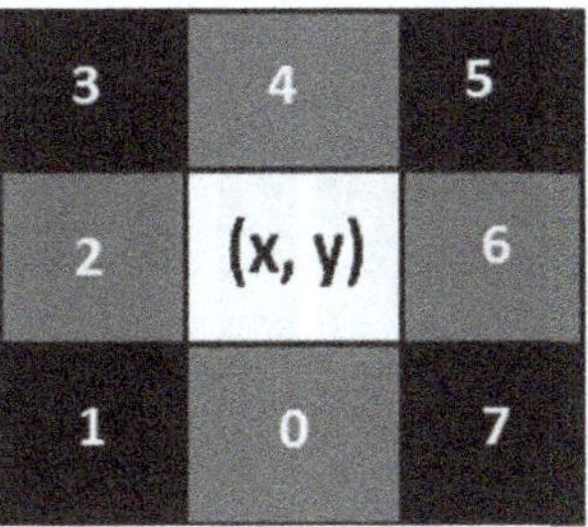

Fig. 2.16 Illustration of determining the pixel neighborhood

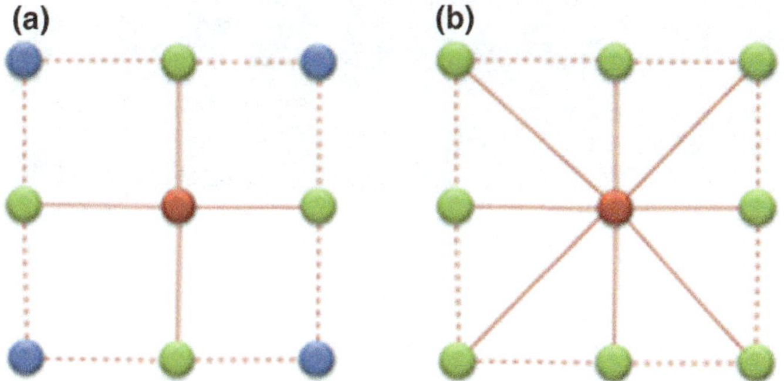

Fig. 2.17 **a**: 4-neighbourhood, **b**: 8-neighbourhood [45]

Neighborhood-based segmentation consists in determining coherent areas, i.e. direct (B-neighbors) and indirect (N-neighbors) neighbors of the selected point.

Figure 2.16 illustrates the method of determining the pixels neighbourhood. For the selected pixel (x, y), in its direct neighbourhood there are pixels with common sides of the pixel. In this case, all the pixels are marked in gray (0, 2, 4, 6). As can be seen, B-neighbourhood pixels are even numbers. In contrast, N-neighbourhood are pixels having common corner with the selected pixels (1, 3, 5, 7). There are two basic types of neighbourhood: 4-neighbourhood and 8-neighbourhood.

Figure 2.17a, b illustrate the neighbourhood. Two points p and q are neighbours if p is included in the 4-neighbourhood of the point q $N_4(q)$ and if q is included in 4-neighbourhood of the point p $N_4(p)$. The procedure is similar in the case of the 8-neighbourhood. The coherence in the sense 4- or 8-neighbourhood can refer as well to the contours and areas [26, 45]. The neighbourhood can be specified as follows

- the Euclidean distance expressed by [26]

$$D = \sqrt{(x_2 - x_1)^2 + (y_2 - y_1)^2},$$
(2.23)

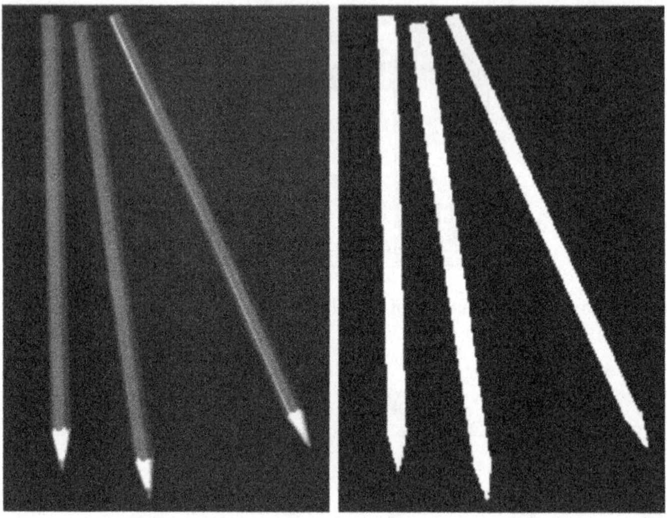

Fig. 2.18 Example of threshold based segmentation [41]

- Distance 4-neighbourhood

$$D = |x_2 - x_1| + |y_2 - y_1|,\tag{2.24}$$

- Distance 8-neighbourhood

$$D = max(|x_2 - x_1|, |y_2 - y_1|).\tag{2.25}$$

The threshold-based segmentation is often used as a preprocessing stage in computer vision. The thresholding operation can be represented by grey value operation g by

$$g(v) = \begin{cases} 0 \; if \; v < t \\ 1 \; if \; v \geq t \end{cases},\tag{2.26}$$

where v is a grey value and t represents the threshold value. The thresholding operation creates the thresholding map which segments the image into two segments, identified by black (1) and white (0) pixels. An example segmentation is presented in Fig. 2.18.

Another type of segmentation is based on merging areas. It divides the image into individual areas by thresholding techniques. The main advantage of this solution is the simplicity of implementation. There is, however, a problem with the selection of thresholds and it usually requires additional filtration logic to remove isolated pixels and operations related to the anti-aliasing or standardisation inside them [31, 60, 62]. An example of such segmentation is presented in Fig. 2.19.

Fig. 2.19 An example of the segmented objects (**b**) from an image (**a**, the image taken from the PASCAL dataset [16])

Use of stereo-vision methods in image segmentation was proposed by Katto et al. [42]. However, their method uses image fusion and a multi-camera system. Their algorithm requires images from four cameras, and they also used k-means clustering. In this case, a certain number of groups is required. In some simulations the mean shift algorithm is used, which does not require setting k parameter [33, 42]. Another interesting algorithm referring to this topic is the work of Toru in which segmentation uses split and merge algorithm [13, 54]. However, in the case of multiple objects, this method does not perform flawlessly.

2.6 Global Features

Global methods extract features from the entire or a part of an image without dividing into more and less significant areas. Examples are histogram-based algorithms such as the histogram of oriented gradients (HOG) or the colour coherence vector (CCV) [12, 47]. In most cases, they generate a constant amount of data which is easier to compare and store, on the other hand, image comparison by histogram-based algorithms gives only a vague similarity for a user. The image area is often divided into even subareas, irrespective of the content of the described area. The features, most often based on histograms, are derived from the mathematical operations performed on all pixels in the region, without defining more or less significant pixels.

The most basic is the colour histogram. The result is a vector of values where the index of the position in the vector corresponds to the colour, while the value itself is the sum of the pixels of that colour. The vector describes the colour of the image. By comparing histograms, we can find similar images by colour similarity, of course, it is possible that they represent a different context.

Gradually, a number of improvements have been made, creating histograms that describe the area more precisely. One of them is the Color Coherence Vector algorithm [47]. In this method, there is a segmentation process, that is, grouping pixels based on the colour value. The method generates a colour histogram, but one colour has two values. The first value is the sum of the pixels of the segments which size was smaller than the threshold value. The second value of the histogram is the sum of the remaining pixels whose segments were larger. This method makes it possible to distinguish, compared to a standard colour histogram, cases such as a black image in white dots, from its counterpart after inversion. In cases where white and black pixels are comparable, colour histograms will be identical, while the improvements allow them to be distinguished efficiently.

Another example of global features is the Histograms of Oriented Gradients [12], in this case, the histogram is generated based on the pixel gradient. The gradient is the first derivative of the pixel values determined in the indicated axis. Most often, the difference between pixels in the vertical and horizontal axes is determined, which gives a two-dimensional gradient vector. The histogram generated by this method makes it possible to distinguish between the image and the gradient and edge of the image. This algorithm is often used for creating specific methods for recognising and classifying images in the learning process.

The advantage of global features is their constant amount of data. Most often the number of parameters is adapted in advance, which is then determined from a fixed number of image areas. A fixed amount of data makes it easier to compare images. Global features describe a picture in a general way. For a human, the effect of the

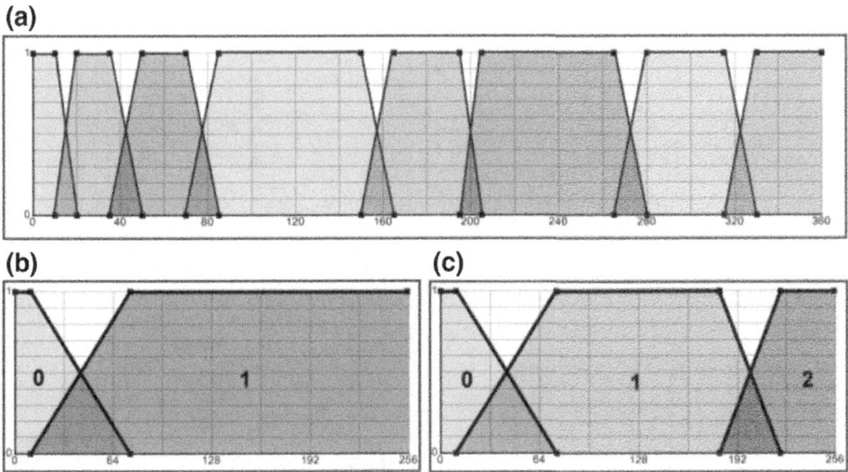

Fig. 2.20 Representations of fuzzy membership functions for the channels of the HSV colour space, respectively: H (**a**), S (**b**), V (**c**) [29]

comparison results at first glance seems to be the same; however, after recognising the images, it turns out that the presented context can be entirely different.

2.6.1 Colour and Edge Directivity CEDD Descriptor

In this section we briefly describe the Color and Edge Directivity Descriptor (CEDD) [9, 29, 35]. CEDD is a global feature descriptor in the form of a histogram obtained by so-called fuzzy-linking. The algorithm uses a two-stage fuzzy system to generate the histogram. A term fuzzy-linking defines that the output histogram is composed of more than one histogram. In the first stage, image blocks in the HSV colour space channels are used to compute a ten-bin histogram. The input channels are described by fuzzy sets as follows [29]:

- the hue (H) channel is divided in 8 fuzzy areas,
- the saturation (S) channel is divided in 2 fuzzy regions,
- the value (V) channel is divided in 3 fuzzy areas.

The membership functions are presented in Fig. 2.20. The output of the fuzzy system is obtained by a set of twenty rules and provides a crisp value [0,1] in order to produce the ten-bin histogram. The histogram bins represent ten preset colours: black, grey, white, red, etc. In the second stage of the fuzzy-linking system, a brightness value of seven colours is computed (without black, grey, white). Similarly to the previous step, S and V channels and image blocks are inputs of the fuzzy system. The output of the second stage is a three-bin histogram of crisp values, which describes the brightness of the colour (light, dark, normal and dark). Both histogram outputs (the

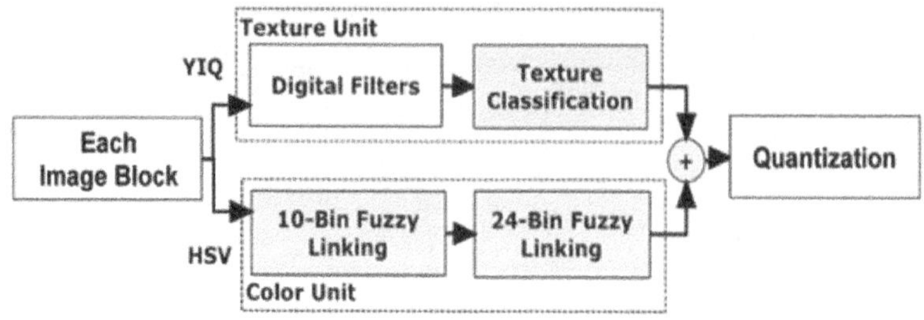

Fig. 2.21 A general schema of computing the CEDD descriptor [29]

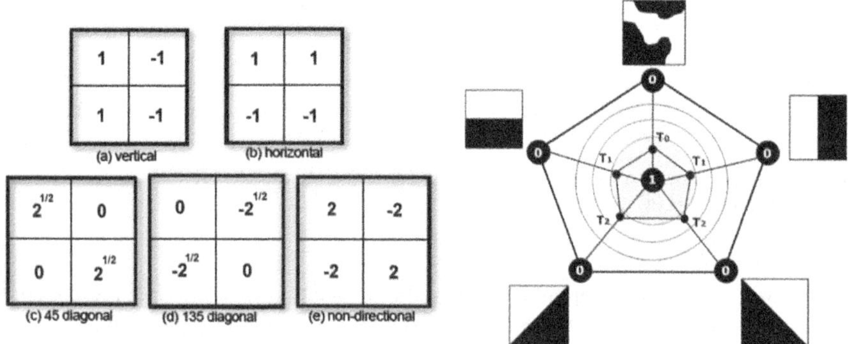

Fig. 2.22 The edge filters used to compute the texture descriptor [29]

first and the second stage) are combined, which allows producing the final 24-bin histogram. Each bin corresponds with colour [29]: (0) Black, (1) Grey, (2) White, (3) Dark Red, (4) Red, (5) Light Red, (6) Dark Orange, (7) Orange, (8) Light Orange, (9) Dark Yellow, (10) Yellow, (11) Light Yellow, (12) Dark Green, (13) Green, (14) Light Green, (15) Dark Cyan, (16) Cyan, (17) Light Cyan, (18) Dark Blue, (19) Blue, (20) Light Blue, (21) Dark Magenta, (22) Magenta, (23) Light Magenta. In parallel to the Colour Unit, a Texture Unit of the Image-Block is computed, which general schema is presented in Fig. 2.21.

In the first step of the Texture Unit, an image block is converted to the YIQ colour space. In order to extract texture information, MPEG-7 digital filters are used. One of these filters is the Edge Histogram Descriptor, which represents five edge types: vertical, horizontal, 45 diagonal, 135 diagonal, and isotropic (Fig. 2.22).

The output of the Texture Unit is a six-bin histogram. When both histograms are computed, we obtain a 144-bin vector for every image block. Then, the vector is normalised and quantised into 8 predefined levels. This is the final step of computing the CEDD descriptor, and now it can be used as a representation of the visual content of the image.

2.7 Summary and Discussion

This chapter presented basic image feature detection methods that are used, among others, to find simple similarity between images. Local feature-based methods try at first to find significant characteristic areas of an image based on Laplacian of Gaussian (LoG) or Difference of Gaussian (DoG) algorithms [25, 64]. Then, they generate a description of their neighbourhood. These methods are more accurate than global ones, on the other hand, they can generate far more description data, and that amount varies per image. Commonly used methods of this kind are SIFT, SURF, ORB, BRIEF, FAST [4, 6, 46, 48, 49]. Local feature methods based on keypoints are efficient in similarity detection between images but less in content recognition.

Global features are extracted from the entire image or from a set of determined areas. They generate a constant amount of data for every image. Finally, the CEDD descriptor was presented. The feature detectors presented in this chapter will be used in the next chapters to retrieve and classify images.

References

1. Al-Amri, S.S., Kalyankar, N.V., et al.: Image segmentation by using threshold techniques (2010). arXiv preprint arXiv:1005.4020
2. Bansal, B., Saini, J.S., Bansal, V., Kaur, G.: Comparison of various edge detection techniques. J. Inf. Oper. Manag. **3**(1), 103–106 (2012)
3. Bao, P., Zhang, L., Wu, X.: Canny edge detection enhancement by scale multiplication. IEEE Trans. Pattern Anal. Mach. Intell. **27**(9), 1485–1490 (2005)
4. Bay, H., Tuytelaars, T., Van Gool, L.: Surf: Speeded up robust features. In: Computer vision–ECCV 2006, pp. 404–417. Springer (2006)
5. Bazarganigilani, M.: Optimized image feature selection using pairwise classifiers. J. Artif. Intell. Soft Comput. Res. **1** (2011)
6. Calonder, M., Lepetit, V., Strecha, C., Fua, P.: Brief: Binary robust independent elementary features. Comput. Vis. ECCV **2010**, 778–792 (2010)
7. Canny, J.: A computational approach to edge detection. Pattern Anal. Mach. Intell. IEEE Trans. **PAMI-8**(6), 679–698 (1986). https://doi.org/10.1109/TPAMI.1986.4767851
8. Chang, Y., Wang, Y., Chen, C., Ricanek, K.: Improved image-based automatic gender classification by feature selection. J. Artif. Intell. Soft Comput. Res. **1**(3), 241–253 (2011)
9. Chatzichristofis, S.A., Boutalis, Y.S.: Cedd: color and edge directivity descriptor: a compact descriptor for image indexing and retrieval. In: International Conference on Computer Vision Systems, pp. 312–322. Springer (2008)
10. Comaniciu, D., Meer, P.: Mean shift analysis and applications. In: The Proceedings of the Seventh IEEE International Conference on Computer Vision, 1999, vol. 2, pp. 1197–1203. IEEE (1999)
11. Comaniciu, D., Meer, P.: Mean shift: a robust approach toward feature space analysis. IEEE Trans. Pattern Anal. Mach. Intell. **24**(5), 603–619 (2002)
12. Dalal, N., Triggs, B.: Histograms of oriented gradients for human detection. In: IEEE Computer Society Conference on Computer Vision and Pattern Recognition, 2005. CVPR 2005, vol. 1, pp. 886–893. IEEE (2005)
13. Damiand, G., Resch, P.: Split-and-merge algorithms defined on topological maps for 3d image segmentation. Gr. Models **65**(1), 149–167 (2003)

14. Derpanis, K.G.: Mean shift clustering. Lecture Notes (2005). http://www.cse.yorku.ca/~kosta/CompVis_Notes/mean_shift.pdf
15. Evans, C.: Notes on the opensurf library. University of Bristol, Technical Report CSTR-09-001, January (2009)
16. Fei-Fei Li, M.A., Ranzato, M.: The pascalobject recognition database collection, unannotated databases - 101 object categories (2009)
17. Georgescu, B., Shimshoni, I., Meer, P.: Mean shift based clustering in high dimensions: a texture classification example. In: Proceedings of Ninth IEEE International Conference on Computer Vision, 2003, pp. 456–463. IEEE (2003)
18. Glantz, R., Pelillo, M., Kropatsch, W.G.: Matching segmentation hierarchies. Int. J. Pattern Recogn. Artif. Intell. **18**(03), 397–424 (2004)
19. Górecki, P., Sopyła, K., Drozda, P.: Ranking by K-means voting algorithm for similar image retrieval. In: International Conference on Artificial Intelligence and Soft Computing, pp. 509–517. Springer (2012)
20. Gould, S., Gao, T., Koller, D.: Region-based segmentation and object detection. In: Advances in Neural Information Processing Systems, pp. 655–663 (2009)
21. Grycuk, R.: Novel visual object descriptor using surf and clustering algorithms. J. Appl. Math. Comput. Mech. **15**(3), 37–46 (2016)
22. Grycuk, R., Gabryel, M., Korytkowski, M., Scherer, R.: Content-based image indexing by data clustering and inverse document frequency. Beyond Databases. Architectures and Structures 2014, Communications in Computer and Information Science, pp. 374–383. Springer, Berlin, Heidelberg (2014)
23. Grycuk, R., Gabryel, M., Korytkowski, M., Scherer, R., Romanowski, J.: Improved digital image segmentation based on stereo vision and mean shift algorithm. In: Parallel Processing and Applied Mathematics 2013, Lecture Notes in Computer Science. Springer Berlin Heidelberg (2014). Manuscript accepted for publication
24. Grycuk, R., Gabryel, M., Korytkowski, M., Scherer, R., Voloshynovskiy, S.: From single image to list of objects based on edge and blob detection. In: International Conference on Artificial Intelligence and Soft Computing, pp. 605–615. Springer (2014)
25. Gunn, S.R.: On the discrete representation of the laplacian of gaussian. Pattern Recogn. **32**(8), 1463–1472 (1999)
26. Haralick, R.M., Shapiro, L.G.: Image segmentation techniques. Comput. Vis. Graph. Image Process. **29**(1), 100–132 (1985)
27. Hare, J.S., Samangooei, S., Lewis, P.H.: Efficient clustering and quantisation of sift features: exploiting characteristics of the sift descriptor and interest region detectors under image inversion. In: Proceedings of the 1st ACM International Conference on Multimedia Retrieval, p. 2. ACM (2011)
28. Hartigan, J.A., Wong, M.A.: Algorithm as 136: a k-means clustering algorithm. J. Royal Stat. Soc Ser. C (Appl. Stat.) **28**(1), 100–108 (1979)
29. Iakovidou, C., Bampis, L., Chatzichristofis, S.A., Boutalis, Y.S., Amanatiadis, A.: Color and edge directivity descriptor on gpgpu. In: 2015 23rd Euromicro International Conference on Parallel, Distributed and Network-Based Processing (PDP), pp. 301–308. IEEE (2015)
30. Jain, R., Kasturi, R., Schunck, B.G.: Machine Vision, vol. 5. McGraw-Hill New York (1995)
31. Jiang, X., Bunke, H.: Edge detection in range images based on scan line approximation. Comput. Vis. Image Underst. **73**(2), 183–199 (1999)
32. Kanungo, T., Mount, D.M., Netanyahu, N.S., Piatko, C.D., Silverman, R., Wu, A.Y.: An efficient k-means clustering algorithm: analysis and implementation. IEEE Trans. Pattern Anal. Mach. Intell. **24**(7), 881–892 (2002)
33. Katto, J., Ohta, M.: Novel algorithms for object extraction using multiple camera inputs. In: Proceedings of International Conference on Image Processing, 1996, vol. 1, pp. 863–866. IEEE (1996)
34. Kirillov, A.: Detecting some simple shapes in images. (2010). http://www.aforgenet.com
35. Kumar, P.P., Aparna, D.K., PhD, V.R.: Compact descriptors for accurate image indexing and retrieval: Fcth and cedd. Int. J. Eng. Res. Technol. (IJERT) **1**, 2278–0181 (2012)

36. Lowe, D.G.: Object recognition from local scale-invariant features. In: The proceedings of the seventh IEEE international conference on Computer vision, 1999, vol. 2, pp. 1150–1157. IEEE (1999)
37. Lowe, D.G.: Distinctive image features from scale-invariant keypoints. Int. J. Comput. Vis. **60**(2), 91–110 (2004)
38. Luo, Y., Duraiswami, R.: Canny edge detection on nvidia cuda. In: IEEE Computer Society Conference on Computer Vision and Pattern Recognition Workshops, CVPRW 2008, pp. 1–8. IEEE (2008)
39. Macedo-Cruz, A., Pajares-Martinsanz, G., Peñas, M.S.: Unsupervised cassification of images in RGB color model and cluster validation techniques. In: IPCV, pp. 526–532 (2010)
40. MacQueen, J., et al.: Some methods for classification and analysis of multivariate observations. In: Proceedings of the fifth Berkeley symposium on mathematical statistics and probability, vol. 1, pp. 281–297. Oakland, CA, USA (1967)
41. Maintz, T.: Digital and Medical Image Processing. Universiteit Utrecht (2005)
42. Marugame, A., Yamada, A., Ohta, M.: Focused object extraction with multiple cameras. Circuits Syst. Video Technol. IEEE Trans. **10**(4), 530–540 (2000)
43. Montazer, G.A., Giveki, D.: Content based image retrieval system using clustered scale invariant feature transforms. Optik-Int. J. Light and Electron. Opt. **126**(18), 1695–1699 (2015)
44. Moon, W.K., Shen, Y.W., Bae, M.S., Huang, C.S., Chen, J.H., Chang, R.F.: Computer-aided tumor detection based on multi-scale blob detection algorithm in automated breast ultrasound images. IEEE Trans. Med. Imag. **32**(7), 1191–1200 (2013)
45. Nakib, A., Najman, L., Talbot, H., Siarry, P.: Application of graph partitioning to image segmentation. Graph Parti., 249–274 (2013)
46. Ng, P.C., Henikoff, S.: Sift: predicting amino acid changes that affect protein function. Nucleic Acid. Res. **31**(13), 3812–3814 (2003)
47. Pass, G., Zabih, R., Miller, J.: Comparing images using color coherence vectors. In: Proceedings of the Fourth ACM International Conference on Multimedia, pp. 65–73. ACM (1997)
48. Rosten, E., Drummond, T.: Machine learning for high-speed corner detection. In: Computer Vision–ECCV 2006, pp. 430–443. Springer (2006)
49. Rublee, E., Rabaud, V., Konolige, K., Bradski, G.: Orb: an efficient alternative to sift or surf. In: 2011 IEEE International Conference on Computer Vision (ICCV), pp. 2564–2571. IEEE (2011)
50. Schreiber, J., Schubert, R., Kuhn, V.: Femur detection in radiographs using template-based registration. In: Bildverarbeitung für die Medizin 2006, pp. 111–115. Springer (2006)
51. Senthilkumaran, N., Rajesh, R.: Edge detection techniques for image segmentation-a survey of soft computing approaches. Int. J. Recent Trends Eng. **1**(2), 250–254 (2009)
52. Shrivakshan, G., Chandrasekar, C., et al.: A comparison of various edge detection techniques used in image processing. IJCSI Int. J. Comput. Sci. Issues **9**(5), 272–276 (2012)
53. Šváb, J., Krajník, T., Faigl, J., Přeučil, L.: Fpga based speeded up robust features. In: IEEE International Conference on Technologies for Practical Robot Applications, 2009. TePRA 2009, pp. 35–41. IEEE (2009)
54. Tamaki, T., Yamamura, T., Ohnishi, N.: Image segmentation and object extraction based on geometric features of regions. In: Electronic Imaging 1999, pp. 937–945. International Society for Optics and Photonics (1998)
55. Tao, D.: The corel database for content based image retrieval (2009)
56. Terriberry, T.B., French, L.M., Helmsen, J.: GPU accelerating speeded-up robust features. In: Proceedings International Symposium on 3D Data Processing, Visualization and Transmission (3DPVT), pp. 355–362. Citeseer (2008)
57. Velmurugan, K., Baboo, L.D.S.S.: Content-based image retrieval using surf and colour moments. Global J. Comput. Sci. Technol. **11**(10) (2011)
58. Wagstaff, K., Cardie, C., Rogers, S., Schrödl, S., et al.: Constrained k-means clustering with background knowledge. ICML **1**, 577–584 (2001)
59. Wang, B., Fan, S.: An improved canny edge detection algorithm. In: Second International Workshop on Computer Science and Engineering, 2009. WCSE 2009. , vol. 1, pp. 497–500. IEEE (2009)

60. Wani, M.A., Batchelor, B.G.: Edge-region-based segmentation of range images. IEEE Trans. Pattern Anal. Mach. Intell. **16**(3), 314–319 (1994). https://doi.org/10.1109/34.276131
61. Wu, M.N., Lin, C.C., Chang, C.C.: Brain tumor detection using color-based k-means clustering segmentation. In: Third International Conference on Intelligent Information Hiding and Multimedia Signal Processing, 2007. IIHMSP 2007, vol. 2, pp. 245–250. IEEE (2007)
62. Wu, Q., Yu, Y.: Two-level image segmentation based on region and edge integration. In: DICTA, pp. 957–966 (2003)
63. Yoon, Y., Ban, K.D., Yoon, H., Kim, J.: Blob extraction based character segmentation method for automatic license plate recognition system. In: 2011 IEEE International Conference on Systems, Man, and Cybernetics (SMC), pp. 2192–2196. IEEE (2011)
64. Young, R.A.: The gaussian derivative model for spatial vision: I. retinal mechanisms. Spat. Vis. **2**(4), 273–293 (1987)

Chapter 3
Image Indexing Techniques

Images are described by various forms of feature descriptors. Especially local invariant features have gained a wide popularity [17, 18, 20, 25, 35]. The most popular local keypoint detectors and descriptors are SURF [1], SIFT [17] or ORB [29]. Such descriptors are long vectors generated in hundreds for a single image; thus it creates a substantial problem with a fast retrieval of identified objects. In order to find similar images to a query image, we need to compare all feature descriptors of all images usually by some distance measures. Such comparison is enormously time-consuming, and there is ongoing worldwide research to speed up the process. However, the current state of the art in the case of high dimensional computer vision applications is not entirely satisfactory. The literature presents many methods and variants, e.g. a voting scheme or histograms of clustered keypoints. They are mostly based on some form of approximate search. One of the solutions to the problem can be descriptor vector hashing. In [6] the authors proposed a locality-sensitive hashing method for the approximate nearest neighbour algorithm. In [25] the authors built a hierarchical quantizer in the form of a tree. Such a tree is a kind of an approximate nearest neighbour algorithm and constitutes a visual dictionary. Recently, the bag-of-features (BoF) approach [9, 27, 35, 43, 45] has gained in popularity. In the BoF method clustered vectors of image features are collected and sorted by the count of occurrence (histograms). There are some modifications of this method, for example, a solution that uses earth movers distance (EMD) presented in [9]. The main problem with the approaches mentioned above is that all individual descriptors or approximations of sets of descriptors presented in the histogram form must be compared. Such calculations are very computationally expensive. Moreover, the BoF approach requires the redesign of the classifier when new visual classes are added to the system. In this chapter, several novel methods to classify and retrieve efficiently various images in large collections.

© Springer Nature Switzerland AG 2020

R. Scherer, *Computer Vision Methods for Fast Image Classification and Retrieval*, Studies in Computational Intelligence 821,
https://doi.org/10.1007/978-3-030-12195-2_3

3.1 Image Classification by Fuzzy Rules

The method presented in this section was presented in [15] and is partly inspired by
the ideas of Viola et al. [40, 42, 46]. They used a modified version of the AdaBoost
algorithm to select the most important features from a large number of elementary
rectangular features similar to Haar basis functions. The selected features are treated
by the authors of [40, 42] as weak classifiers for the content-based image retrieval
task, mainly for images containing faces. Contrary to the previous authors, who
developed CBIR systems based on boosting techniques, in the approach proposed
in this section we use the original version AdaBoost algorithm to choose the most
important local features. A wide variety of local and global visual feature descriptors
(e.g. SURF, SIFT or ERB, see Sect. 2.1) can be used, and the method applies to a
broader class of images (not only face images). Moreover, incorporating new visual
classes in the system requires only adding new fuzzy rules to the rule base without
restructuring the existing rule base.

We propose a novel approach to use fuzzy logic and fuzzy rules as the adjustable
representation of visual feature clusters. Fuzzy logic [30, 31, 33, 34] is a very convenient method for describing partial membership to a set. We combine fuzzy logic and
boosting meta-learning to choose the most representative set of image features for
every considered class of objects. In each step, we randomly choose one feature from
a set of positive images taking into consideration feature weights computed using
the AdaBoost algorithm. This feature constitutes a base to build a weak classifier.
The weak classifier is given in the form of a fuzzy rule, and the selected feature is
a base to determine the initial parameters of the fuzzy rule. In the learning process,
the weak classifiers are adjusted to fit positive image examples. This approach could
be beneficial for the search based on the image content in a set of complex graphical
objects in a database. The main contribution and novelty of the system presented in
this section are: a novel method for automatic building a fuzzy rule base for image
classification based on local features, an efficient technique for fast classification of
images, a method for automatic search of the most salient local features for a given
class of images, and a flexible image indexing system. In the next subsection, a new
method of creating a weak classifier ensemble as a fuzzy rule base is presented.

The algorithm presented in this section uses the AdaBoost algorithm which is the
most popular boosting method [32]. The algorithm described here is designed for
binary classification. Let us denote the l-th learning vector by $\mathbf{z}^l = [x_1^l, ..., x_n^l, y^l]$,
$l = 1...L$ is the number of a vector in the learning sequence, n is the dimension of
input vector $\mathbf{x}^l$, and y^l is the learning class label. Weights D^l assigned to learning
vectors, have to fulfill the following conditions

$$\text{(i)}\ 0 < D^l < 1,$$
$$\text{(ii)}\ \sum_{l=1}^{L} D^l = 1. \tag{3.1}$$

The weight D^l is the information how well classifiers were learned in consecutive steps of an algorithm for a given input vector $\mathbf{x}^l$. Vector $\mathbf{D}$ for all input vectors is initialized according to the following equation

$$D_t^l = \frac{1}{L}, \quad \text{for } t = 0, \ldots, T, \tag{3.2}$$

where t is the number of a boosting iteration (and a number of a classifier in the ensemble). Let $\{h_t(\mathbf{x}) : t = 1, \ldots, T\}$ denotes a set of hypotheses obtained in consecutive steps t of the algorithm being described. For simplicity we limit our problem to a binary classification (dichotomy), i.e., $y \in \{-1, 1\}$ or $h_t(\mathbf{x}) = \pm 1$. Similarly to learning vectors weights, we assign a weight c_t for every hypothesis, such that

$$\begin{aligned} &\text{(i)} \sum_{t=1}^{T} c_t = 1 \,, \\ &\text{(ii)} \ c_t > 0 \,. \end{aligned} \tag{3.3}$$

Now in the AdaBoost algorithm we repeat steps 1–4 for $t = 1, \ldots, T$:
1. Create hypothesis h_t and train it with a data set with respect to a distribution d_t for input vectors.
2. Compute the classification error ε_t of a trained classifier h_t according to the formula

$$\varepsilon_t = \sum_{l=1}^{m} D_t^l(z^l) I(h_t(\mathbf{x}^l) \neq y^l), \tag{3.4}$$

where I is the indicator function

$$I(a \neq b) = \begin{cases} 1 \text{ if } a \neq b \\ 0 \text{ if } a = b \end{cases}. \tag{3.5}$$

If $\varepsilon_t = 0$ or $\varepsilon_t \geq 0.5$, stop the algorithm.
3. Compute the value

$$\alpha_t = 0.5 \ln \frac{1 - \varepsilon_t}{\varepsilon_t}. \tag{3.6}$$

4. Modify weights for learning vectors according to the formula

$$D_{t+1}(\mathbf{z}^l) = \frac{D_t(\mathbf{z}^l) \exp\{-\alpha_t \mathbf{I}(h_t(\mathbf{x}^l) = y^l)\}}{N_t}, \tag{3.7}$$

where N_t is a constant such that $\sum_{l=1}^{m} D_{t+1}(\mathbf{z}^l) = 1$. To compute the overall output of the ensemble of classifiers trained by the AdaBoost algorithm, the following formula is used

$$f(\mathbf{x}) = \sum_{t=1}^{T} c_t h_t(\mathbf{x}), \tag{3.8}$$

where

$$c_t = \frac{\alpha_t}{\sum_{t=1}^{T} \alpha_t} \tag{3.9}$$

is the classifier importance for a given training set, $h_t(\mathbf{x})$ is the response of the hypothesis t on the basis of feature vector $\mathbf{x} = [x_1, \ldots, x_n]$. The coefficient c_t value is computed on the basis of the classifier error and can be interpreted as the measure of classification accuracy of the given classifier. Moreover, the assumption (3.1) should be met. As we see, the AdaBoost algorithm is a meta-learning algorithm and does not determine the way of learning for classifiers in the ensemble.

3.1.1 Boosting-Generated Simple Fuzzy Classifiers

Here we try to find the most representative fuzzy rules for a given class ω_c, $c = 1, \ldots, V$, of visual objects and to fast classify query images afterwards. This section describes the learning process, i.e. generating fuzzy rules from a set of examples. The algorithm uses the boosting meta-learning to generate a suitable number of weak classifiers. The classifiers feature space $\mathbb{R}^N$ consists of elements x_n, $n = 1, \ldots, N$. For example, in the case of using the standard SIFT descriptors (see Sect. 2.1.1), $N = 128$.

In each step, we randomly choose one local feature from the set of positive images according to its boosting weight. Then, we search for similar feature vectors from all positive images. Using these similar features, we construct one fuzzy rule. Undoubtedly, it is impossible to find exactly the same features in all images from the same class; thus we search for feature vectors which are similar to the feature picked randomly in the current step. This is one of the reasons for using fuzzy sets and fuzzy logic. The rules have the following form

$$\begin{aligned} R_t^c : \text{ IF } & x_1 \text{ is } G_{1,t}^c \text{ AND } x_2 \text{ is } G_{2,t}^c \text{ AND} \ldots \\ & \ldots \text{ AND } x_{128} \text{ is } G_{128,t}^c \text{ THEN image } i \in \omega_c(\beta_t^c) \text{ ,} \end{aligned} \tag{3.10}$$

where $t = 1, \ldots, T^c$ is the rule number in the current run of boosting, T^c is the number of rules for the class ω_c and β_t^c is the importance of the classifier, designed to classify objects from the class ω_c, created in the tth boosting run. The weak classifiers (3.10) consist of fuzzy sets with Gaussian membership functions

$$G_{n,t}^c(x) = e^{-\left(\frac{x - m_{n,t}^c}{\sigma_{n,t}^c}\right)^2}, \tag{3.11}$$

where $m_{n,t}^c$ is the center of the Gaussian function (3.11) and $\sigma_{n,t}^c$ is its width. For the clarity of the presentation this section describes generating the ensemble of weak classifiers for a class ω_c; thus the class index c will be omitted.

Let I be the number of all images in the learning set, divided into two sets: positive images and negative images, having, respectively, I_{pos} and I_{neg} elements. Obviously $I = I_{pos} + I_{neg}$. Positive images belong to a class ω_c that we train our classifier with. For every image from these two sets, we determine local features, for example, local interest points using, e.g. SIFT or SURF algorithms. The points are represented by descriptors, and we operate on two sets of vectors: positive descriptors $\{\mathbf{p}^i; i = 1, \ldots, L_{pos}\}$ and negative ones $\{\mathbf{n}^j; j = 1, \ldots, L_{neg}\}$. In the case of the standard SIFT algorithm, each vector $\mathbf{p}^i$ and $\mathbf{n}^j$ consists of 128 real values. Let v^i be the number of keypoint vectors in the ith positive image, let u^j be the number of keypoint vectors in the jth negative image. Then, the total number of learning vectors is given by

$$L = \sum_{i=1}^{I_{pos}} v^i + \sum_{j=1}^{I_{neg}} u^j, \tag{3.12}$$

where $L = L_{pos} + L_{neg}$. According to the AdaBoost algorithm, we have to assign a weight to each keypoint in the learning set. When creating new classifiers, the weights are used to indicate keypoints which were difficult to handle. At the start of the algorithm, all the weights have the same, normalised values

$$D_1^l = \frac{1}{L} \text{ for } l = 1, \ldots, L. \tag{3.13}$$

Let us define matrices $\mathbf{P}_t$ and $\mathbf{N}_t$ constituting the learning set

$$\mathbf{P}_t = \begin{bmatrix} \mathbf{p}^1 & D_t^1 \\ \vdots & \vdots \\ \mathbf{p}^{L_{pos}} & D_t^{L_{pos}} \end{bmatrix} = \begin{bmatrix} p_1^1, \ldots, p_N^1 & D_t^1 \\ \vdots & \vdots \\ p_1^{L_{pos}}, \ldots, p_N^{L_{pos}} & D_t^{L_{pos}} \end{bmatrix}, \tag{3.14}$$

$$\mathbf{N}_t = \begin{bmatrix} \mathbf{n}^1 & D^1 \\ \vdots & \vdots \\ \mathbf{n}^{L_{neg}} & D_t^{L_{neg}} \end{bmatrix} = \begin{bmatrix} n_1^1, \ldots, p_N^1 & D^1 \\ \vdots & \vdots \\ n_1^{L_{neg}}, \ldots, p_N^{L_{neg}} & D_t^{L_{neg}} \end{bmatrix}. \tag{3.15}$$

The learning process consists in creating T simple classifiers (weak learners in boosting terminology) in the form of fuzzy rules (3.10). After each run t, $t = 1, \ldots, T$, of the proposed algorithm, one fuzzy rule R_t is obtained. The process of building a single fuzzy classifier is presented below.

1. Randomly choose one vector $\mathbf{p}^r$, $1 \leq r \leq L_{pos}$ from positive samples using normalized distribution of elements $D_t^1, \ldots, D_t^{L_{pos}}$ in matrix (3.14). This drawn vec-

tor becomes a basis to generate a new classifier and the learning set weights contribute to the probability of choosing a keypoint.

2. For each image from the positive image set find the feature vector which is nearest to $\mathbf{p}^r$ (for example according to the Euclidean distance) and store this vector in matrix $\mathbf{M}_t$ of the size $I_p \times N$. Every row represents one feature from a different image v_i, $i = 1, \ldots, I_{pos}$, and no image occurs more than once

$$
\mathbf{M}_t = \begin{bmatrix} \tilde{p}_{t,1}^1 & \cdots & \tilde{p}_{t,N}^1 \\ \vdots & \cdots & \vdots \\ \tilde{p}_{t,1}^j & \ddots & \tilde{j}_{t,N}^j \\ \vdots & \cdots & \vdots \\ \tilde{p}_{t,1}^{I_{pos}} & \cdots & \tilde{p}_{t,N}^{I_{pos}} \end{bmatrix},
\tag{3.16}
$$

Each vector $\left[\tilde{p}_{t,1}^j \cdots \tilde{p}_{t,N}^j \right]$, $j = 1, \ldots, I_{pos}$, in matrix (3.16) contains one visual descriptor from the set $\{\mathbf{p}^i; i = 1, \ldots, L_{pos}\}$. For example, in view of descriptions (3.14) and (3.12), the first row in matrix (3.16) is one of the rows of the following matrix

$$
\begin{bmatrix} p_1^1, \ldots, p_N^1 \\ \vdots & \vdots \\ p_1^{v_1}, \ldots, p_N^{v_1} \end{bmatrix},
\tag{3.17}
$$

where v_1 is the number of feature vectors in the first positive image.

3. In this step a weak classifier is built, i.e. we find centres and widths of Gaussian functions which are membership functions of fuzzy sets in a fuzzy rule (3.10).

  a. Compute the absolute value $d_{t,n}$ as the difference between the smallest and the highest values in each column of the matrix (3.16)

$$
d_{t,n} = |\min_{i=1,\ldots,I_p} p_n^i - \max_{i=1,\ldots,I_p} p_n^i|
\tag{3.18}
$$

  where $n = 1, \ldots, N$. Compute the center of fuzzy Gaussian membership function (3.11) $m_{t,n}$ in the following way

$$
m_{t,n} = \max_{i=1,\ldots,I_p} p_n^i - \frac{d_{t,n}}{2}.
\tag{3.19}
$$

Now we have to find the widths of these fuzzy set membership functions. We have to assume that for all real arguments in the range of $\left[m_{t,n} - \frac{d_{t,n}}{2}; m_{t,n} + \frac{d_{t,n}}{2} \right]$ the Gaussian function (fuzzy set membership function) values should satisfy $G_{n,t}(x) \geq 0.5$. Only in this situation do we activate the fuzzy rule. As we assume that $G_{n,t}(x)$ is at least 0.5 to activate a fuzzy rule, using simple substitution $x = m_{t,n} - \frac{d_{t,n}}{2}$, we obtain the relationship for $\sigma_{t,n}$

$$\sigma_{t,n} = \frac{d_{t,n}}{2\sqrt{-\ln(0.5)}} \tag{3.20}$$

Finally, we have to calculate the values $m_{t,n}$ and $\sigma_{n,t}$ for every element of the nth column of matrix (3.16), thus we have to repeat the above steps for all N dimensions. In this way, we obtain N Gaussian membership functions of N fuzzy sets. Of course, we can label them using fuzzy linguistic expressions such as 'small', 'large' etc., but for the time being we mark them only in a mathematical sense by $G_{n,t}$, where $n, n = 1, \ldots, N$, is the index associated with feature vector elements and t means the fuzzy rule number.

   b. Using values obtained in point a) we can construct a fuzzy rule which creates a fuzzy classifier (3.10).

4. Now we have to evaluate the quality of the classifier obtained in step 3. We do this using the standard AdaBoost algorithm [32]. Let us determine the activation level of the rule R_t which is computed by a t-norm of all fuzzy sets membership function values

$$f_t(\bar{\mathbf{x}}) = \mathop{T}_{n=1}^{N} G_{n,t}(\bar{x}_n), \tag{3.21}$$

where $\bar{\mathbf{x}} = [\bar{x}_1, \ldots, \bar{x}_N]$ is a vector of values of linguistic variables $x_1, \ldots, x_N$. In the case of minimum t-norm formula (3.21) becomes

$$f_t(\bar{\mathbf{x}}) = \min_{n=1}^{N} G_{n,t}(\bar{x}_n). \tag{3.22}$$

As a current run of the AdaBoost is for a given class ω_c, we can treat the problem as a binary classification (dichotomy) i.e. $y^l = 1$ for descriptors of positive images and $y^l = 0$ for descriptors of negative images. Then the fuzzy classifier decision is computed by

$$h_t(\bar{\mathbf{x}}^l) = \begin{cases} 1 & \text{if } f_t(\bar{\mathbf{x}}^l) \geq \frac{1}{2} \\ 0 & \text{otherwise} \end{cases}. \tag{3.23}$$

For all the keypoints stored in matrices $\mathbf{P}_t$ and $\mathbf{N}_t$ we calculate new weights D_t^l. To this end, we compute the error of classifier (3.23) for all $L = L_{pos} + L_{neg}$ descriptors of all positive and negative images

$$\varepsilon_t = \sum_{l=1}^{L} D_t^l I(h_t(\bar{\mathbf{x}}^l) \neq y^l), \tag{3.24}$$

where I is the indicator function

$$I(a \neq b) = \begin{cases} 1 & \text{if } a \neq b \\ 0 & \text{if } a = b \end{cases}. \tag{3.25}$$

If $\varepsilon_t = 0$ or $\varepsilon_t > 0.5$, we finish the training stage. If not, we compute new weights:

$$\alpha_t = 0.5 \ln \frac{1 - \varepsilon_t}{\varepsilon_t}. \tag{3.26}$$

$$D_{t+1}^l = \frac{D_t^l \exp\{-\alpha_t I(h_t(\bar{\mathbf{x}}^l) = y^l)\}}{C}, \tag{3.27}$$

where C is a constant such that $\sum_{l=1}^{L} D_{t+1}^l = 1$. Finally classifier importance is determined by

$$\beta_t = \frac{\alpha_t}{\sum_{t=1}^{T} \alpha_t}. \tag{3.28}$$

It should be noted that the classifier importance (3.28) is needed to compute the overall response of the boosting ensemble for the query image, which will be described in detail in the next section. The concept of 'word' used in the BoW method [9, 27, 35, 45] corresponds to a fuzzy rule in the presented method, which in the case of the SIFT application, consists of 128 Gaussian functions. The next section will describe a classification of a new query image by the ensemble.

3.1.2   Classification of a Query Image

The boosting procedure described in the previous section should be executed for every visual object class ω_c, $c = 1, \ldots, V$, thus after the learning procedure, we obtain a set of V strong classifiers. Let us assume that we have a new query image and an associated set of u visual features represented by matrix $\mathbf{Q}$

$$\mathbf{Q} = \begin{bmatrix} \mathbf{q}^1 \\ \mathbf{q}^2 \\ \vdots \\ \mathbf{q}^u \end{bmatrix} = \begin{bmatrix} q_1^1 \cdots q_N^1 \\ q_1^2 \cdots q_N^2 \\ \vdots \\ q_1^u \cdots q_N^u \end{bmatrix}. \tag{3.29}$$

Let us determine the value of

$$F_t(\mathbf{Q}) = \mathop{S}_{j=1}^{u} \left(\mathop{T}_{n=1}^{N} G_{n,t}(q_n^j) \right), \tag{3.30}$$

where S and T are t-norm and t-conorm, respectively (see [31]). To compute the overall output of the ensemble of classifiers designed in Sect. 3.1.1, for each class ω_c we sum weak classifiers outputs (3.30) taking into consideration their importance (3.28), i.e.

$$H^c(\mathbf{Q}) = \sum_{t=1}^{T^c} \beta_t F_t(\mathbf{Q}). \tag{3.31}$$

Eventually, we assign a class label to the query image in the following way

$$f(\mathbf{Q}) = \arg\max_{c=1,\ldots,V} H^{*c}(\mathbf{Q}). \tag{3.32}$$

In formulas (3.31) and (3.32) we restored class label index c, which had been removed at the beginning of Sect. 3.1.1. In formula (3.30) t-norm and t-conorm can be chosen as min and max operators, respectively.

3.1.3 Experiments

The goal of the experiments was to evaluate the proposed approach and compare it with the state-of-the-art method in terms of accuracy and speed. We tested the proposed method on three classes of visual objects taken from the PASCAL Visual Object Classes (VOC) dataset [8], namely: Bus, Cat and Train. Examples of such visual objects are presented in Fig. 3.1. We divided these three classes of objects into learning and testing examples. The testing set consists of 15% images from the whole dataset. Before the learning procedure, we generated local keypoint vectors for all images from the Pascal VOC dataset using the SIFT algorithm. These 128-element vectors were stored in separate files for every image in the dataset. Each file contained hundreds of vectors, depending on the complexity of the image.

Fig. 3.1 Examples of objects from Bus, Cat and Train class taken from PASCAL Visual Object Classes (VOC) dataset

All simulations were performed on a Hyper-V virtual machine with MS Windows Operating System (8 GB RAM, Intel Xeon X5650, 2.67 GHz). We determined the time and quality of classification. The testing set contained only images that had never been presented to the system during the learning process.

The proposed method was implemented using C# language. For the learning process needs, we built an extra set of examples (negative examples) for each considered class of objects. The negative examples were picked by random from other classes. We chose only the most representative objects for considered classes, whereas for the testing purposes we chose various kinds of images from a considered class.

We have compared our results with the bag-of-features image representation model combined with the Support Vector Machine (SVM) classification based on the Chi-Square kernel. The BoF algorithm is currently one of the most popular algorithms in computer vision, and it was run five times for various dictionary sizes: 200, 250, 300, 350 and 400 words. Dictionaries for the BoF were created using C++ language, based on the OpenCV Library [4]. The BoF experiments were performed on the same set of objects as the experiments for the method proposed in this section. The results of the BoF and SVM classification, both learning and testing, are presented in Table 3.1. In the BoF algorithm the learning process (dictionary generation) is run globally for all classes, thus column LT is empty in Table 3.1 for each class. In Table 3.2 we depict experiment results of the method described in Sects. 3.1.1 and 3.1.2. The BoF combined with the SVM algorithm achieved the best overall classification accuracy for three classes for the dictionary of size 350 (Table 3.1), which was approximately worse by 2% than in the case of the proposed method. Moreover, the learning and classification time for the proposed method is considerably shorter than the BoF-SVM (182.117 s vs 246.48 s), which is better, respectively by 35% and 32%. The dictionaries of other sizes performed slightly worse than the dictionary of size 350. It can be seen that our method gives a better classification accuracy and the time of learning and testing process is shorter than in the case of the SVM algorithm. It should also be emphasised that our method has an extra advantage; namely, we can add a new class of visual objects to the existing system by just adding new rules. In the BoF we have to recreate the whole dictionary. The most time-consuming part of the bag-of-features classification is the SVM learning.

3.1.4 Conclusions

We proposed a new approach to fast image classification. Our approach, which works by repeatedly creating fuzzy rules based on most salient image features, has shown promising results on a real-world dataset. Despite its simplicity, it outperformed the bag-of-features method in terms of accuracy and speed. It demonstrates the following advantages:

- the method is relatively accurate in visual object classification,
- learning and classification is very fast,

Table 3.1 Results of the learning and testing processes for the BoF and SVM algorithms (CQ—Classification Quality (%), LT—Learning time (s), TT—Testing time (s)). The learning time is given only as the overall time for all classes

	CQ (%)	LT	TT
Dictionary size: 200			
Buses	70.59		3.532
Cats	100.00		5.199
Trains	41.17		4.833
Total	41.18	195.570	13.564
Dictionary size: 250			
Buses	70.59		3.627
Cats	88.24		5.858
Trains	35.29		5.177
Total	64.71	208.241	14.662
Dictionary size: 300			
Buses	76.47		3.678
Cats	88.24		5.734
Trains	41.18		5.134
Total	68.63	213.317	14.546
Dictionary size: 350			
Buses	70.59		3.696
Cats	94.12		5.862
Trains	52.94		5.436
Total	72.55	246.480	14.994
Dictionary size: 400			
Buses	70.59		4.116
Cats	88.24		6.136
Trains	52.94		5.344
Total	70.59	265.469	15.596

Table 3.2 Classification accuracy and time of the learning and testing processes obtained by the method proposed in this section (CQ—Classification Quality (%), LT—Learning time (s), TT—Testing time (s)). The learning time is given only as the overall time for all classes

	Positive learning samples	Negative Learning samples	Classif. accuracy on testing set (%)	Learning time (s)	Testing time (s)
Buses	76	17	82.35		3.236
Cats	82	17	76.47		4.495
Trains	73	17	64.71		3.593
Total	231	51	74.51	182.117	11.324

- expanding the system knowledge is efficient as adding new visual classes to the system requires the generation of new fuzzy rules whereas in the case of bag-of-features it requires new dictionary generation and re-learning of classifiers.

The method also demonstrates potential in terms of possibility to expand it in order to incorporate different features or different meta-learning algorithms.

It should be noted that the system can work with virtually any type of fuzzy membership functions, e.g. triangular or bell-shape. Moreover, various types of t-norm can be used in the algorithm, but the application of the minimum t-norm leads to faster computation than in the case of other t-norms.

3.2 Salient Object Detector and Descriptor by Edge Crawler

One of the problems in computer vision is object detection, i.e. determining the region of interest (ROI). The salient objects need to be extracted, and then, some mathematical transformations are performed in order to obtain their description. The accuracy of the results depends heavily on the homogeneity of the objects. Unfortunately, many image datasets and real-world imagery contain images of objects with a non-homogeneous background. This section presents an algorithm for image description based on edge detection (in the presented case Canny, described in Sect. 2.2) and local keypoints (in our case SURF, see Sect. 2.1.2). We developed a method which extracts only essential features of the object. The proposed algorithm takes an input image and returns a list of angle-keypoint histograms.

In the first step, we perform SURF keypoint detection and description on all the images from the dataset. Afterwards, the keypoints are clustered by the k-means algorithm. The clustering is needed to obtain centroids for keypoint histograms described later in this section. The number of clusters has to be set in advance by some experimental procedure if we use k-means clustering.

Then, we detect edges in each input image by the Canny algorithm. Unfortunately, the crawler described later should be executed on continuous edges (see Fig. 2.12b). To eliminate this issue, we have to link broken edges (see Fig 2.12c). Then, we proceed to the crawler algorithm. The crawler traverses through pixels. The algorithm finishes when the first pixel is reached again or all points are processed. The detailed description of the algorithm is shown in Algorithm 2 and an exemplary route in Fig. 3.2 (for more details see [10]). The algorithm is executed to surround the object found by the edge detection. The proposed algorithm moves between pixels and calculates angles between them. The algorithm starts with finding the first pixel of an edge. Each found salient pixel neighbourhood is used to determine the consecutive pixel. In the case of multiple pixels in the proximity, we choose the first pixel clockwise (see Fig. 3.4). The position of the branch is stored in the form of a stack (last in, first out) and labelled as visited. In the next step, we calculate the angle between the next pixel and the previous pixel (see Fig. 3.5). We quantize angles to the following values:

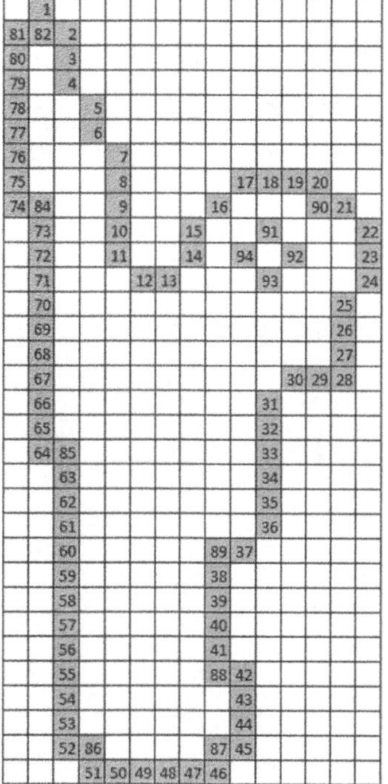

Fig. 3.2 Example of the crawler route. It is used for locating regions of interest

$\phi \in \{45, 90, 135, 180, 225, 270, 315\}$. The quantization is sufficient to describe the shape and allows for faster descriptor comparison. Afterwards, the SURF keypoints are examined if they belong (are located in the object's ROI) to any of the detected objects. The keypoints which are outside the objects are removed. An example is presented in Fig. 3.6.

After the keypoints are assigned to the objects a histogram of keypoints is generated for every object. It is constructed based on keypoint cluster centroids for all the images and determines how many keypoints of the current object are clustered to a given cluster (e.g. *cluster* 1 contains 4 keypoints). The number of bins in the keypoint histogram depends on the number of centroids for all keypoints and is set or determined by some procedure at the beginning of the indexation stage.

In the last step of the presented method, we concatenate previously created histograms (angle and keypoint) to obtain one histogram that describes a given object. It should be added, that the histogram is normalized. The algorithm steps are presented in pseudo-code (Algorithm 3). In Fig. 3.3 we present the block diagram with the data structures of the algorithm.

The presented method is a fast and robust visual feature extractor. It allows to immune the descriptor to change of scale or rotation, which provides very good results in all the experiments we performed. The immunity comes from the robust keypoint descriptors and the angle descriptor. The algorithm was used as a feature

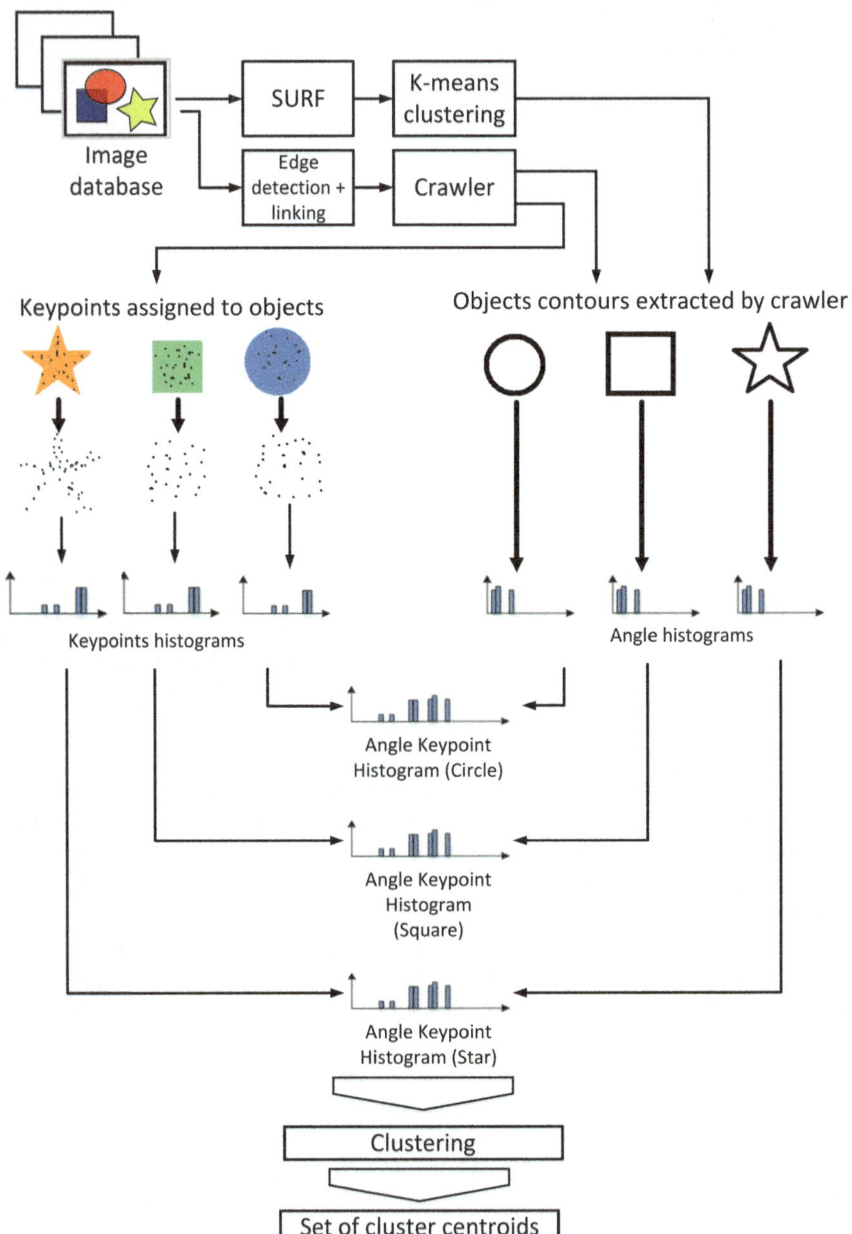

Fig. 3.3 Image database indexing stage with resulting data structures

INPUT: $ObjectContours$
OUTPUT: $AngleHistogram$
$EdgeDetectedImage := CannyEdgeDetection(InputImage);$
$EdgeDetectedImage := EdgeLinking(EdgeDetectedImage);$
$CurrentPixel := FindPositionOfNearestPixel(EdgeDetectedImage);$
$IsCrawlingCompleted := false;$
while $IsCrawlingCompleted = true$ **do**
    $NextPixel = FindNextPixel();$
    while $NextPixel! = NULL$ **do**
        $VisitedPixels.Add(CurrentPixel);$
        $PrevPixel := CurrentPixel;$
        $CurrentPixel := NextPixel;$
        $NextPixel := FindNextPixel();$
        $CurrentPixel.Angle := Angle(PrevPixel, CurrentPixel, NextPixel);$
        if $VisitedPixels.Contains(NextPixel) = true$ **then**
            $NextPixel := NULL;$
            if $Branches.Count = 0$ **then**
                $IsCrawlingCompleted := true;$
            end
        end
    end
    if $Branches.Count > 0$ **then**
        $PrevPixel := NULL;$
        $CurrentPixel := Branches.Last();$
    end
    else if $Branches.Count = 0$ **then**
        $IsCrawlingCompleted := true;$
    end
end

Algorithm 2: Crawler algorithm steps

extraction and indexing step in the image retrieval system presented in the next section (Fig. 3.7).

3.2.1 System for Content-Based Image Retrieval

In this section we present the system we used for evaluating the descriptor proposed earlier. The system is described in Fig. 3.8. All images in the database have to be indexed initially by the algorithm proposed in the previous section. Namely, we have to

- compute SURF descriptors for all images and cluster them,
- perform edge detection and linking,
- run crawler and generate the angle histograms,
- select salient keypoints,
- generate angle-keypoint histograms.

INPUT: *InputImages*
OUTPUT: *AngleKeypointHistograms*
DetectedKeypoints := *SurfDetector(InputImages)*;
ClusterKeypoints(DetectedKeypoints);
foreach *InputImage* ∈ *InputImages* **do**
   EdgeDetectImg := *Canny(InputImage)*;
   EdgeDetectImg := *EdgeLinking(EdgeDetectImg)*;
   RunCrawler(EdgeDetectedImage);
   ObjectsList := *ExtractObjects()*;
   foreach *Obj* ∈ *ObjectsList* **do**
      ObjAngleHist := *CreateAngleHist(Obj)*
      ObjKeypoints := *ExtractKeyPoints(Obj, DetectedKeypoints)*;
      ObjKeypintsHist := *CreateKeyPointHist(ObjKeypoints, Obj)*;
      AngleKeyPointHist := *Concat(ObjAngleHist, ObjKeypintsHist)*;
      AngleKeypointHistograms.Add(AngleKeyPointHist);
   end
end

Algorithm 3: The image indexing algorithm

We used k-means clustering, thus we had to choose in advance the number of groups. A higher number of groups provided better accuracy at the expense of the speed. It is possible to use other clustering methods such as mean shift or cam shift which set the number of clusters automatically. Our experiments with them showed that they are very sensitive to unbalanced data. Obviously, local features can be computed by different methods such as SIFT, ORB, FAST or FREAK: Fast Retina Keypoint. The same applies to the edge detection step.

After generating descriptors and cluster them, the system is ready to accept query images. Of course, the query image has to be processed by the same method as the rest of images in the database. Moreover, the processing must be performed using the same parameters to provide the same descriptor size. In the case of the presented system, to index all the images in the dataset, we have to:

- compute SURF descriptors for all images,
- perform edge detection and linking,
- run crawler and generate the angle histograms,
- select salient keypoints,
- assign the salient keypoints to groups from the image database,
- generate angle-keypoint histograms.

Then, we check the similarity of the generated histograms for the query image to the centroids of groups from the database.

Sometimes there is a necessity to add a new set of images to the database. If the new images are from the same domain it is enough to compute their features with the procedure used in the image retrieval. However, if the new images come from a substantially different domain the whole existing database must be reindexed with the procedure from Sect. 3.2.

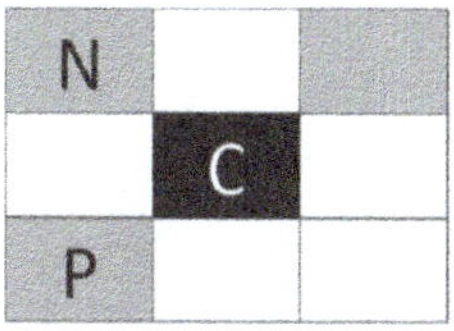

Fig. 3.4 Determining the next pixel in the crawler route, P is the previous pixel, C is the current one and N determines the next pixel

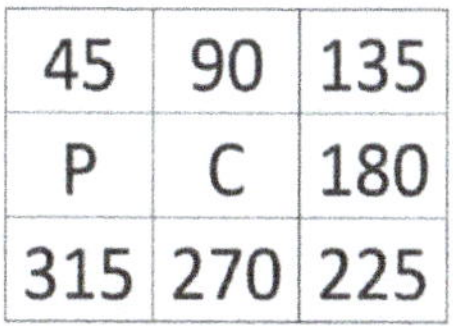

45	90	135
P	C	180
315	270	225

Fig. 3.5 Calculating the angle of the next pixel, P is the previous pixel and C is the current one

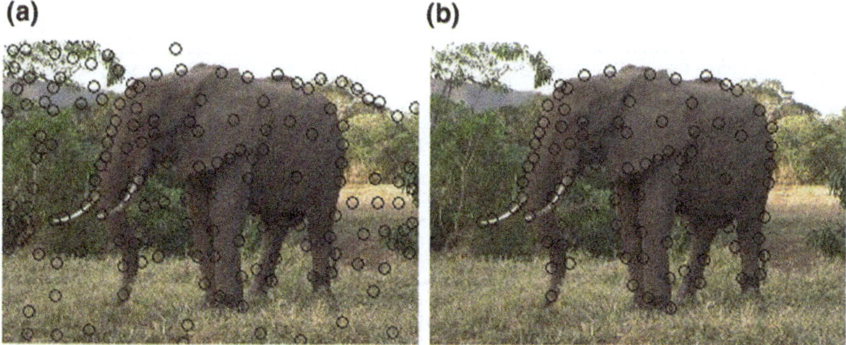

Fig. 3.6 Removing points that lie outside objects. Figure B consists of important keypoints and points from the background are discarded as a result of the proposed method

3.2.2 *Experimental Results*

We carried out experiments using software described in [11] on two image datasets:

- The PASCAL Object Recognition Database Collection of 101 Unannotated Object Categories [8],
- The COREL Database for Content-based Image Retrieval [37–39].

We used all classes in the datasets and every class was divided into two sets of, respectively, 90% training images and (10%) query (test) images. The performance of the proposed method was evaluated with *Precision* and *Recall* measures [5, 41]. They are computed using:

- AI—appropriate images which should be returned,
- RI—returned images by the system,

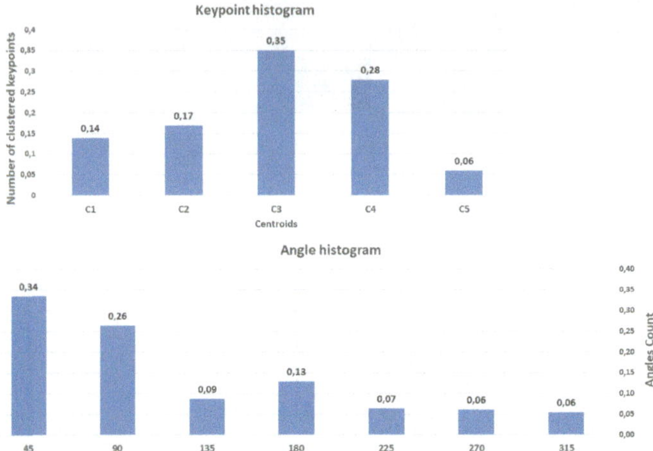

Fig. 3.7 The concatenated histogram. It is composed of angle and keypoint histograms. In the figure, it was split in two and presented one beneath the other

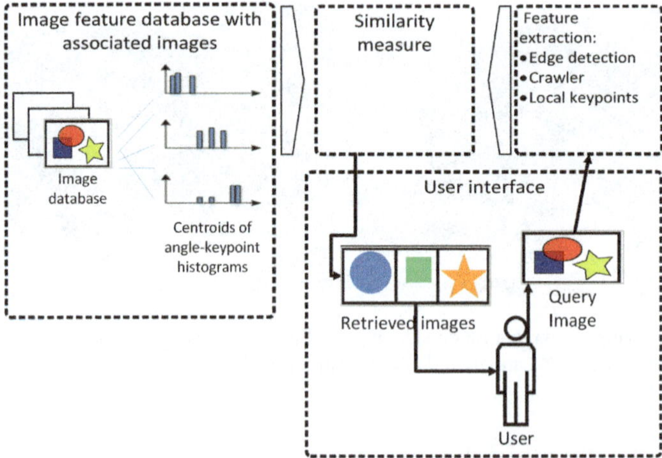

Fig. 3.8 Content-based image retrieval architecture utilizing the method proposed in Sect. 3.2

- Rai—properly returned images (intersection of AI and RI),
- Iri—improperly returned images,
- anr—proper not returned images,
- inr—improper not returned images,

and $precision$ and $recall$ are defined [19] (Fig. 3.9)

$$precision = \frac{|rai|}{|rai + iri|},$$

(3.33)

Table 3.3 Experiment results for the proposed algorithm, performed on the Pascal dataset. Due to lack of space, we present only a part of all queries from various classes, although Avg. Precision is calculated for all query images in the test set

Id(Class)	RI	AI	rai	iri	anr	Precision	Recall
5(helicopter)	78	72	54	24	18	69	75
4(mandolin)	38	36	31	7	5	82	86
8(accordion)	48	45	35	13	10	73	78
3(crocodile)	44	41	33	11	8	75	80
7(soccer-ball)	57	53	42	15	11	74	79
6(pizza)	46	44	29	17	15	63	66
2(butterfly)	81	74	61	20	13	75	82
11(mandolin)	39	36	31	8	5	79	86
12(accordion)	48	45	34	14	11	71	76
13(crocodile)	43	41	35	8	6	81	85
14(soccer-ball)	58	53	42	16	11	72	79
16(butterfly)	77	74	61	16	13	79	82
10(helicopter)	78	72	53	25	19	68	74
17(mandolin)	38	36	33	5	3	87	92
19(accordion)	47	45	34	13	11	72	76
11 32(mayfly)	35	33	29	6	4	83	88
35(soccer-ball)	57	53	42	15	11	74	79
29(Faces)	381	353	325	56	28	85	92
38(helicopter)	76	72	54	22	18	71	75
Avg. precision						73.42	

$$recall = \frac{|rai|}{|rai + anr|}. \tag{3.34}$$

Table 3.3 shows the measures of retrieved images of chosen classes the Pascal dataset. As can be seen, the results are satisfying and they are better than our previous results obtained in [10]. Due to lack of space, we present only random classes from the whole dataset. The presented results prove that the proposed approach can be useful in CBIR techniques.

Fig. 3.9 Performance measures diagram [12]

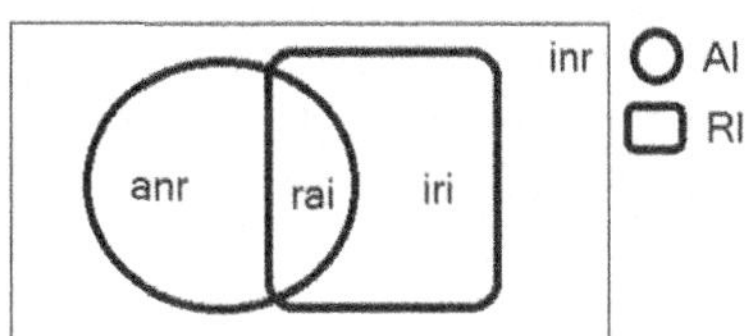

Table 3.4 Experiment results for the proposed algorithm, performed on the Corel dataset. Due to lack of space, we present only a part of all queries from various classes, although Avg. Precision is calculated for all query images in the test set

Id(Class)	RI	AI	rai	iri	anr	Precision	Recall
1(art)	98	90	76	22	14	78	84
2(art)	98	90	73	25	17	74	81
3(art)	95	90	73	22	17	77	81
4(art)	97	90	74	23	16	76	82
9(art)	96	90	78	18	12	81	87
10(art)	97	90	77	20	13	79	86
11(antiques)	96	90	76	20	14	79	84
12(antiques)	98	90	77	21	13	79	86
13(antiques)	97	90	73	24	17	75	81
21(cybr)	199	190	165	34	25	83	87
22(cybr)	200	190	173	27	17	86	91
23(cybr)	205	190	177	28	13	86	93
31(dino)	96	90	82	14	8	85	91
32(dino)	97	90	82	15	8	85	91
33(dino)	98	90	80	18	10	82	89
41(mural)	96	90	69	27	21	72	77
42(mural)	94	90	62	32	28	66	69
43(mural)	96	90	69	27	21	72	77
50(mural)	95	90	63	32	27	66	70
Avg. Precision						78.58	

In Table 3.4 factors for randomly chosen classes of retrieved images from the Corel dataset are presented. As can be seen in example query 23(cybr) most images were correctly retrieved (177). Only 28 of them are improperly recognized. The *precision* value for this experiment equals 86 and *recall* equals 93. The average precision equals 78.58 which is a very good result on the Pascal dataset. Figures 3.10 and 3.11 show the average precision for each class for both datasets.

The time performance of the proposed method is determined by the size of images, hardware and the algorithm parameters, e.g. in the case of SURF keypoints on the value of the minHessian coefficient. We used one 12-core Intel Xeon E5-2620 2.0Ghz processor and the indexation step (without keypoint generation and edge detection) for the Pascal VOC dataset took approximately 10 s. The image retrieval step during experiments (i.e. for 10% of all images) took, respectively, 8 min. 56 s. for the Corel dataset and 20 min. 36 s. for the Pascal VOC dataset.

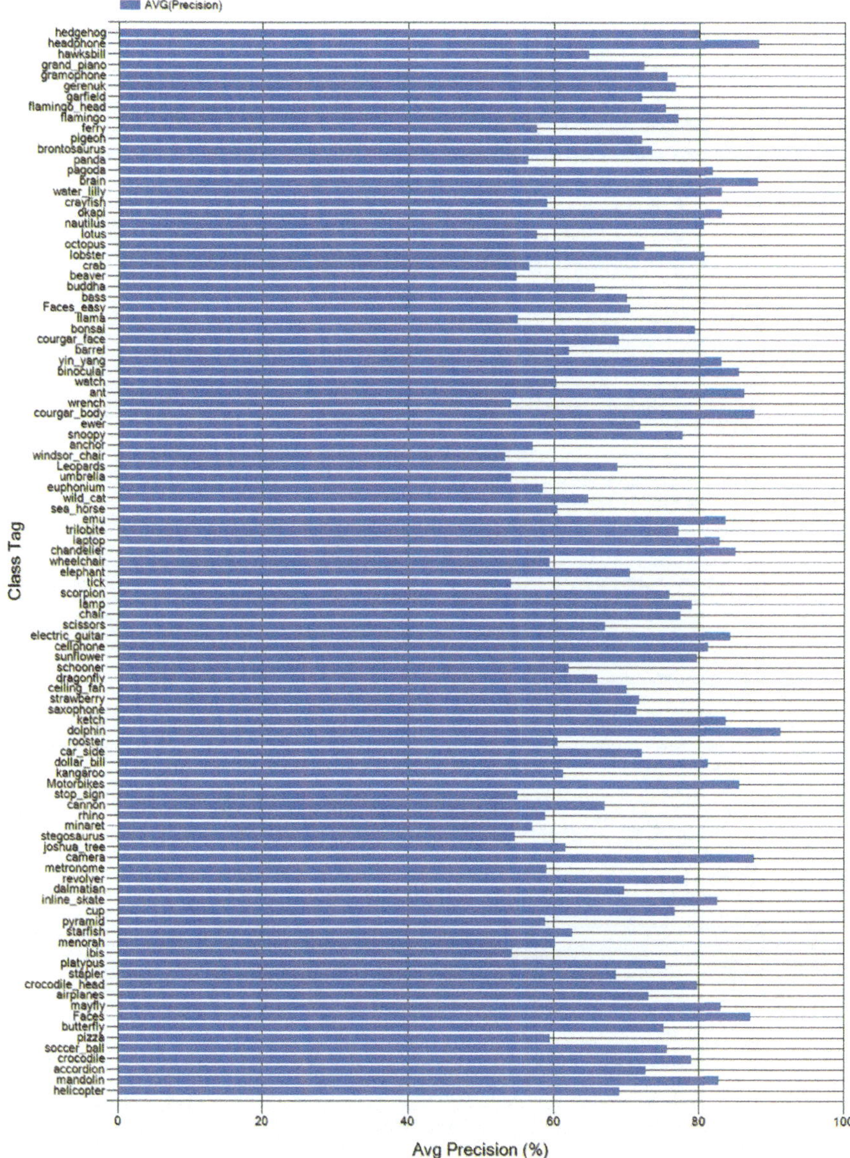

Fig. 3.10 Average Precision for each image class for the proposed algorithm, performed on the Pascal dataset

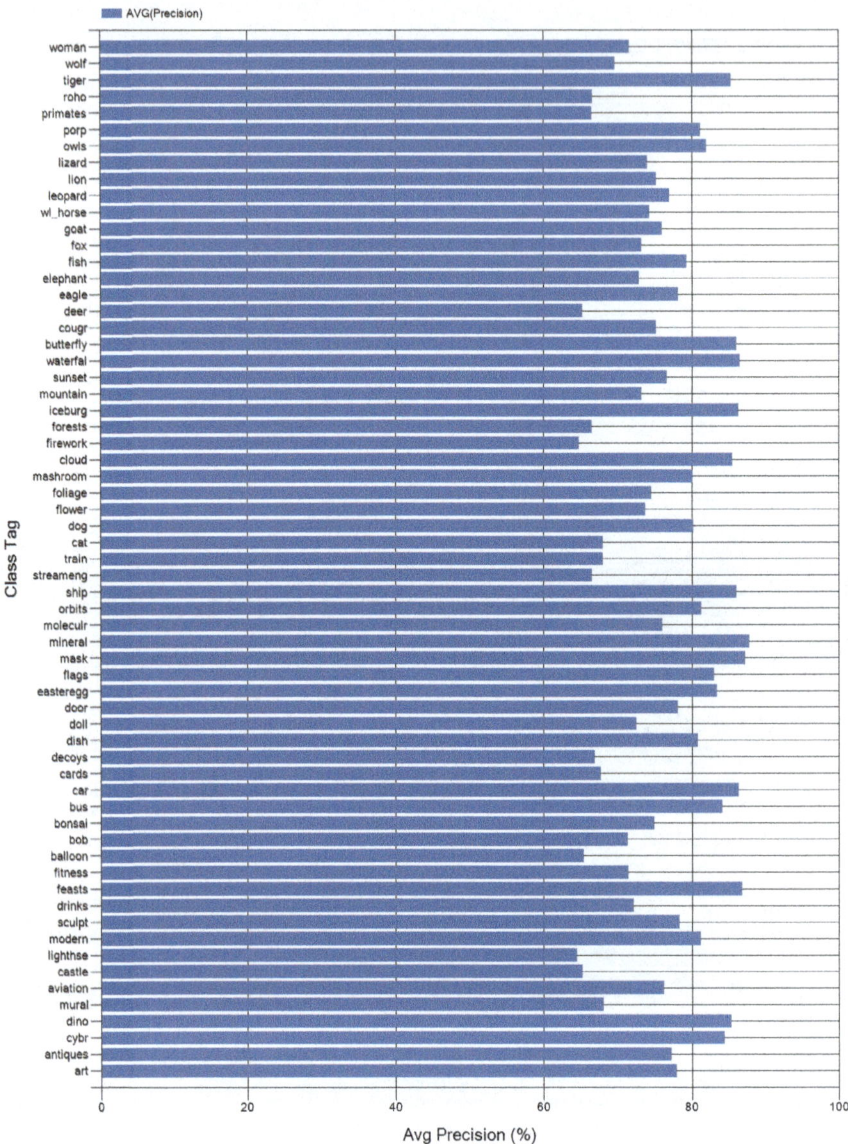

Fig. 3.11 Average Precision for each image class for the proposed algorithm, performed on the Corel dataset

3.2.3 Conclusions

The new framework for content-based image retrieval based on local image features presented in this section relays on the crawler algorithm to extract salient local features. As an example, we use SURF image keypoints that are extracted for each detected object. Of course, it is possible to use other local image feature extractors and descriptors. In the end, we build a histogram of salient keypoints and the object outline angles. Such a descriptor is relatively fast to compute and immune to scale change or rotation. The method provides the angle-keypoint histogram as an object descriptor; thus the comparison in the indexation phase (see Fig. 3.8) is relatively fast and straightforward.

The performed experiments showed that the introduced set of algorithms is efficient in terms of speed and accuracy. The approach can be applied to various vision systems.

Although slightly less accurate, the proposed algorithm is much faster than the solutions based on convolutional neural networks. The retrieval speed is at least ten times faster in the case of the presented framework.

3.3 Fast Two-Level Image Indexing

In order to improve the process of descriptor comparison in large image collections, this section proposes a method based on dual-hashing of the keypoint descriptors [21]. The primary goal of our approach is to group similar descriptors and to represent them by two integer values acting as the main index and the sub-index. Using two integer values, calculated as hashes from the SURF descriptors, we resolve the problem of descriptor values between two single hash values.

As aforementioned, the SURF algorithm (Sect. 2.12) is a feature extractor designed to search the exact similar content of the entire image or its fragments. To this end, the method tries to find unique points of the image, describe their surrounding context, and finally, to compare with a set of keypoints from another image. In SURF, keypoints are generated similarly to blob detection; thus, depending on the image complexity, the method can produce a large number of points. Algorithms based on keypoints do not distinguish more or less significant points, and we do not know which point will match a point in another image. In the case of perspective change of the image, a lot of points disappear, and new ones appear. Thus, such algorithms operate on sets of points extracted from images and try to find best-fit pairs. In most cases, several pairs of thousands compared is enough to determine the similarity between images. For example, for a small image, the method extracted 588 and 762 keypoints per image, but only 50 best pairs were used to determine the similarity between the images.

A similarity between keypoints can be determined by a simple distance measure between descriptor vectors. In this section, we use the normalised 64-dimensional

SURF descriptor version; thus the distance between descriptors varies from 0 (best fit) to 1 (worst fit).

The simplest way to compare two sets of descriptors is to check each combination between the sets, and that makes a considerable number of descriptors to compare. In most cases, when an algorithm is used for tracing an object, it compares multiple times similar sets of points. That process could be sped up by using a method that can learn or reduce the set of points to most matched. A different case is when we have to find similarity in a collection of images. For example, to query by image content, to find an object or to stitch a set of photos into one. In that case, when we need to search the collection multiple times, and on demand, we could prepare searched image collection before, by keeping the extracted features in a file. Then, the method does not need to access the image content and can operate only on the extracted features. The number of extracted points per image makes a huge collection of points that must be compared in a single query.

The different problem is that descriptors are related to only the local neighbourhood of the keypoint. Thus it is impossible to order or select more or less significant values from the descriptor array. All of them are equally important and independent, and in many cases, single or more values could have noise. Because of that, we compare them by determining the distance between the descriptor vectors. This is a problem in the case when we need to reduce the descriptor to a hash code when a single value peak could change the entire hash code.

The last problem, that must be considered with keypoint-based methods, is that descriptors are responsible for only a small fragment of the image content, and in some cases, this fragment will be similar to other fragments. Thus, some keypoints will not be unique, even in the space of its image and could match to multiple points on the compared image. Therefore, by this fact, after descriptor comparison, the method must check the correlation between matched points location, on both images to confirm the proper match.

3.3.1   Hash Generation

Each of the two hash values is represented by a 32-bit integer and is divided into 16 groups of bits (Fig. 3.12). Each group represents a value between 0 and 3 encoded on 2 bits. To determine the bit group value, the method first adds respective elements of the descriptor vector according to Eq. 3.35, where the result value is always between 0 and 4

$$s_n = \frac{1}{4} \sum_{i=0}^{4} desc_{16i+n}^2 , \tag{3.35}$$

where $n = 0, \ldots, 15$ is the group number, $desc_n$ is nth SURF descriptor vector element. After that, the method rounds s_n to an integer by the floor function (Eq. 3.36) to compute the main index elements $g1_n$.

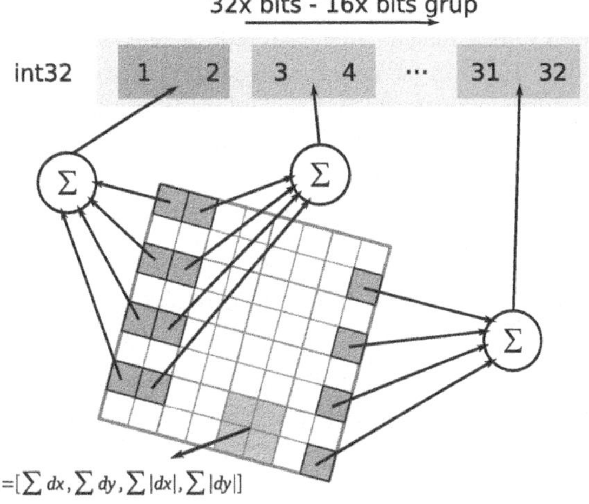

Fig. 3.12 The structure of the proposed hash value divided into sixteen bit-groups computed from the SURF descriptor vector elements

$$g1_n = \lfloor s_n + 0.5 \rfloor \tag{3.36}$$

In order to compute the sub-index hash value $g2_n$, the method takes the value $g1_n$ of the main index hash group and decreases it if the fractional part is less 0.3, increases if more than 0.7, otherwise leaves it unchanged

$$g2_n = \begin{cases} g1_n - 1 & \text{if } \{s_n\} < 0.3 \\ g1_n + 1 & \text{if } \{s_n\} > 0.7 \\ g1_n \end{cases} \tag{3.37}$$

In the proposed method, we used two bits per group. This is due to our assumption, that if the distance between normalised descriptors is larger than 0.25, then we consider them as entirely different. Thus, when we sum not whole but certain values of a descriptor (in our method only four per bit group), then the sum will oscillate between 0 and 1. Next, if we divide this sum by 4, we obtain a value between 0 and 4 that can be stored on 2 bytes. Thus, when we compare a bit group between descriptors, then if their values are the same we consider them as possibly similar, otherwise as entirely different. However, in some cases, when the value after division is for example 1.45 and is rounded to 1 and from a different descriptor is 1.55 rounded to 2, then we have a problem when in a hash we lose the information that they might be similar. Because of this problem, our method uses the second hash value that in this example will take value 2 for the main descriptor and 1 for the sub-descriptor. By using the second hash, we do not lose information regarding the close similarity between descriptors.

Finally, we have two hash values per each descriptor. The main hash encodes the exact information of descriptors that corresponds to the fragments of the descriptor under the bit group. The second hash value describes the correlation of the descriptor to other close hash values. Thus, in the case when the descriptor value is between two hash values, then the first hash is the nearest value and second is the correlation to the other. Then, in the case of image comparison, the method compares the descriptor group together as the first-second index and second-first index. Thus, thanks to these two hash codes the proposed method can compare descriptors of different hash codes but of very similar value.

3.3.2 Structure of the Proposed Descriptor Index

In our method, we use an index representation stored in memory and a file. In the memory index, the method keeps descriptors that have been just extracted from an image, e.g. for the current query image. This type of structure presents a class diagram in Fig. 3.14. In the memory representation, the method used map container that allows fast access to an element by its index value. The method use Node elements to organise descriptors in smaller sets identified by the hash value. Keypoints are stored in the SecondNode object as an unordered list. The SecondNode is a sub-element of the FirstNode and they are stored in a map structure identified by the second hash. SecondNode is common for multiple numbers of the descriptor of the same hash values. Moreover, at last, FirstNode objects are stored in the root element in a map structure indexed by the first hash value of related descriptors.

File representation shown in Fig. 3.14 is different from the memory one. In the file, each of elements is saved one by one, so at first, the method writes the first hash corresponding to the FirstNode, then a subset of the second hash (SecondNode) and, at last, their descriptor list. The files are easily read and write from the memory structure. To reduce memory usage, keypoints and descriptors are changed. In keypoint, we reduce parameters initially stored by int32 to minimal int8 (rotation, size), and int16 (position). The descriptor is also reduced, from the float representation to uint8. Thus, after this process, we reduce keypoint and descriptor sizes from 272B to 70B. This reduction gains importance with a large number of keypoints, that are extracted from an image and stored in a file.

Thus, when we add an image to the index, the proposed method extracts keypoints and descriptors and create memory representation. Then, the method saves the structure corresponding to the image to a file (Fig. 3.13).

3.3.3 Index Search Process

In our method, keypoints and descriptors are stored in files. Each file is related to an indexed image. In the search process, we do not load this files to memory but check files on the fly.

In the memory, we keep descriptors from the query image because we use only one query image in contrast to a lot of indexed images. Thus, in the search process, only

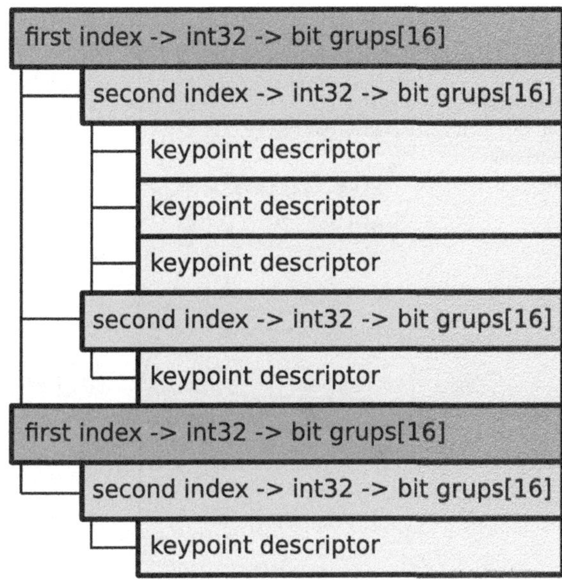

Fig. 3.13 Example of the structure of the proposed two-level descriptor index

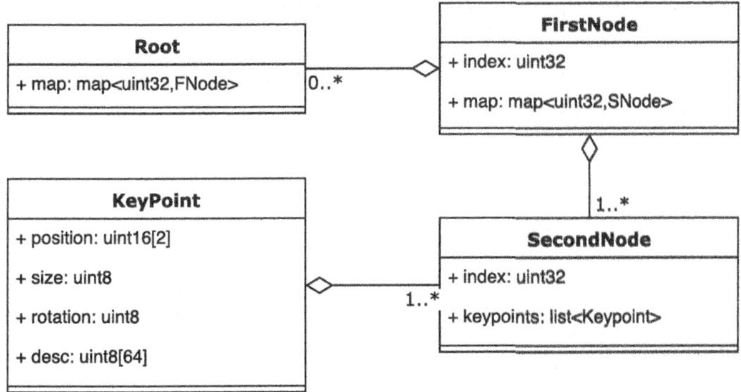

Fig. 3.14 The class diagram presents the structure of the proposed descriptor index in memory

the query image is compared multiple times, while each file only once. In our method, we use a kind of reverse data mining approach. The active index is created from the query image. Then it is multiple times searched and compared with descriptors from files.

The main advantage of the method is that we do not compare the entire combination of descriptors between sets, but only descriptors that might be similar. This approach significantly reduces the number of comparisons. We indicate small groups of descriptors for full comparison by their indexed hash, that is very fast in consideration to description comparison that is based on the vector distance. This relation

Fig. 3.15 Example of descriptor comparison between nodes of two indexes. Numbers represent examples of hash values divided into bit groups that represent each group of descriptors

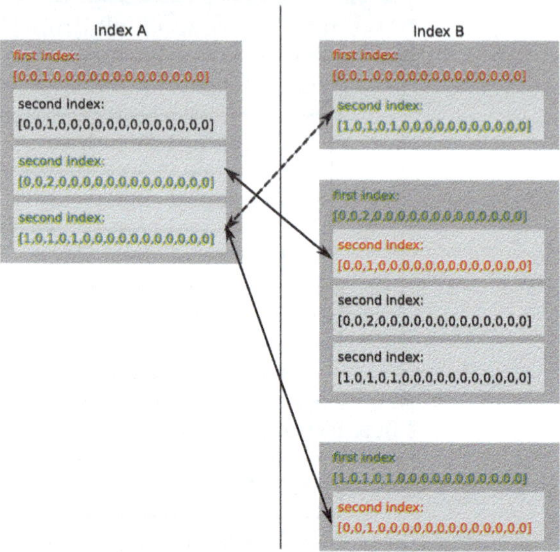

between the groups of descriptors is presented in Fig. 3.15. Thus, we compare two combinations of hash values. The first is exact (main index -> sub-index) marked by the dashed line and second opposite (sub-index -> main index) marked by the solid lines. By this approach, we compare only the descriptors which distance might be less than 0.25; other descriptors are omitted.

3.3.4 *Experimental Results*

We performed two experiments implemented in C++ on AMD A6-6310 CPU with 16GB of memory in Linux Ubuntu 17. The first one on a small test set of images, that contained similar content in a different perspective to test similarity search. In the second experiment, we used the PASCAL Visual Object Classes (VOC) dataset [8] to check the performance of the proposed approach.

The results of a single search of the first experiment are presented in Fig. 3.16, where the first image from the left is the query image, and the next seven images are the results of matching, i.e. best-matched images. In this experiment, the searched images where rated by a number of descriptor pairs, which distance was smaller than 0.25. As we can see, the method can select similarity based only on the number of similar descriptors, but in real cases the method must also check the correlation between keypoint positions, to confirm the similarity of content. Then, the method discards the results from the second row of Fig. 3.16.

In this experiment, we noticed the problem with keypoints that describe a small image area with simple content. They create a large number of random pairs. These

Table 3.5 Index search results per image from Fig. 3.16

Image	1	2	3	4	5	6	7
Descriptors	1429	1361	1475	1075	1039	955	1232
First index	97	101	102	98	90	87	94
Second index	715	696	703	621	595	572	612
Descriptor match	1513	575	557	510	491	484	433

pairs dominate in results even if create only single pair based on the best fit, what marginalises more proper and unique pairs. To avoid this problem, we add a parameter of punishment to each descriptor, that count number of combination where distance is less than 0.25. Then the punishment parameter of the entire image decreases its resultant rate of similarity. After that improvement, the method obtains more proper results as is shown in Fig. 3.16. Also, in the case of examination of keypoint correlation, avoiding these pairs of descriptors speeds up the process because most of them represent wrong matching.

The results of the first experiment are presented in two tables. Table 3.5 contains information about image index stored in files for images (Fig. 3.16) started from the second image. The second row describes the number of descriptors extracted from the image. The number of the descriptors depends on the image content complexity and the image size. In our experiment to avoid the problem of a large set of descriptors, we resize each image to the maximal size of 1000x1000 pixels. The next row describes the number of the first nodes, as we see each node groups have about 13 descriptors. The third row represents the number of the second nodes that is about 7 per the first node and contains on average two descriptors. The last row describes the number of matched descriptors.

Table 3.6 presents the results of search example from Fig. 3.16. The entire set contains 79 images. Time of search was 182 ms and depends on the system configuration, and does not include the time spend on descriptor extraction from the query image. Most of this time, the method spend on descriptor distance comparison. The third row contains the number of the second nodes that were compared in this single query. Each node contains descriptors that also were compared. The fourth row presents the number of all descriptors combination that must be considered to check in the case of the brute force method. The fifth row presents the number of descriptors that were compared. As we see, in the proposed method we need only about 0.07% of the entire combination to compare, and from this number, about 43% descriptors were matched. Thus in conclusion, in the case of the brute force method, we must compare each combination of descriptors, but in this example, only about 0.03% of them is matched. The presented method saves a lot of computational resources.

The second experiment was conducted on the PASCAL Visual Object Classes (VOC) dataset [8], which contains about 16 thousand images. At first, the method extracts descriptors from each image and stores them in files. Then for each file,

Fig. 3.16 Results of the first experiment. The first image on the left is the query image. Other images are results of the search

Table 3.6 Summary of index search from Fig. 3.16

Parameter	Value
Image index size	79
Search time	182 ms
Second groups compared	5956
Descriptors combinations	101 414 000
Descriptors compared	79 057
Percent of compared to combination	0.077%
Descriptors matched	34 640
Percent of matched to combination	0.034%
Percent of matched to compared	43.816%

the method loads descriptor files to memory and queries the entire set. Thus, in the experiment, each image was compared with each other. In Table 3.7 we present the summary of index creation of this experiment divided on minimal, average, and maximal achieved values. From the single image, the method extracts about six hundred keypoints (that number oscillates between seven and three thousand). The average ratio between the number of descriptors of the first and the last node is similar to the first experimental results. In Table 3.8 we present the summary of the search experiment. The second row shows the number of descriptor combination. As we see, these numbers are huge comparing to the number of images. In row three and four, we see that the proposed indexing scheme needs to compare much fewer descriptors and that about 28% of them is matched.

Table 3.7 Experimental results of indexing of the PASCAL Visual Object Classes (VOC) dataset

Parameter	Min. value	Avg. value	Max. value
Images	–	16,125	–
Descriptors	7	597	3 314
Second groups	6	371	753
First groups	5	74	124
Descriptors per second group	1	1.44	4.33
Descriptors per first group	1	6.94	21.74
Second per first group	1	4.75	8.63

Table 3.8 Results of index search experiment for the VOC dataset

Parameter	Min. value	Avg. value	Max. value
Images	–	16 125	–
Descriptors combination	12 138 468	7 674 154 032	14 625 496 000
Descriptors compared	2 329	5 728 754	46 091 539
Descriptors match	216	1 669 547	36 548 916
Index search time	4 342 ms	10 762 ms	30 681 ms
Compared/combination	0.0082%	0.0821%	0.7542%
Matched/compared	0.14%	28.73%	84.32%
Matched/combination	0.0002%	0.0282%	0.6090%

3.3.5   Conclusions

We presented a method that is able to index a large image dataset for retrieval purposes efficiently. It is based on the SURF keypoint descriptors. The method can significantly reduce the number of descriptor comparisons to 0.03% of all comparison combinations. The algorithm allows to store indexed visual features in files, and search them on demand on the fly, without loading them to memory, which simplifies work with larger sets of images. Descriptors represented by simple integer values also reduce data usage. Data usage does not exceed 100 KB for 1000 descriptors. Of course, this amount of data might be significant when compared to image file sizes in the case of using small images. The disadvantage of the presented approach is that the method operates on a relatively large set of features per image. Keypoint-based algorithms generate many points, and most of them will never be associated with keypoints from other images. Of course, the proposed approach can be adapted to different types of visual features, where we need to compare elements by distance like it is in the family of keypoint descriptors.

3.4 Image Colour Descriptor

This section presents an image descriptor [22] based on color spatial distribution for image similarity comparison. It is similar to methods based on HOG and spatial pyramid but in contrast to them operates on colours and colour directions instead of oriented gradients. The presented method assumes using two types of descriptors. The first one is used to describe segments of similar colour and the second sub-descriptor describes connections between different adjacent segments. By this means we gain the ability to describe image parts in a more complex way as is in the case of the histogram of oriented gradients (HOG) algorithm but more general as is in the case of keypoint-based methods such as SURF or SIFT. Moreover, in comparison to the keypoint-based methods, the proposed descriptor is less memory demanding and needs only a single step of image data processing. Descriptor comparing is more complicated but allows for descriptor ordering and for avoiding some unnecessary comparison operations.

In most global feature-based methods, image description is too simple and cannot provide satisfactory results for accurate image comparison or classification. Colour histogram-based methods in most cases bring only a vague similarity. On the other hand, they are efficient in comparing large sets of images. In the case of local feature-based algorithms, we face different problems: difficult comparison and classification due to an irregular keypoints distribution over the image and descriptors that describe only a small scrap of space around the keypoint. Local keypoint descriptors represent blobs and corners of the image which not adequately represent the real, abstract image content.

To address the aforementioned problems, the research presented in this Section [22] focuses on several goals:

- Describing image in a more precise way than in the case of global, histogram-based features, and in a more readable way for the human. Color histogram-based methods describe only colours what does not allow guessing the image content for a human. Similarly, local features are hard to recognise by humans as they describe many single elements of an image. This problem is shown in Fig. 3.17.
- Obtaining a small number of generated descriptors per image. Local feature-based methods generate hundreds of keypoints (vectors) per image. In many cases, keypoints are located in the same object (structure). The presented method creates a single descriptor for the entire structure that replaces multiple keypoints. For example, in the case of a triangle presented in Fig. 3.17, SURF detects three keypoints, whereas in the proposed method the entire triangle is represented by a single descriptor.
- Creating a descriptor that will be able to be sorted for comparison speed-up. Most local and global feature descriptors do not distinguish more or less essential values. Each descriptor parameter corresponds to a relevant element of space around keypoints or for a single colour in the case of histograms. They need to be compared

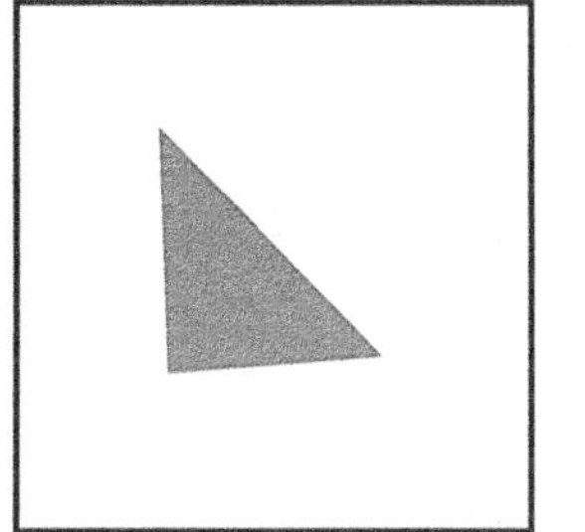
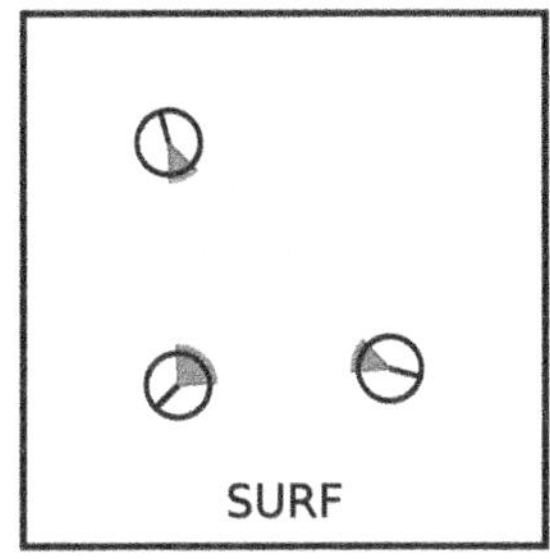

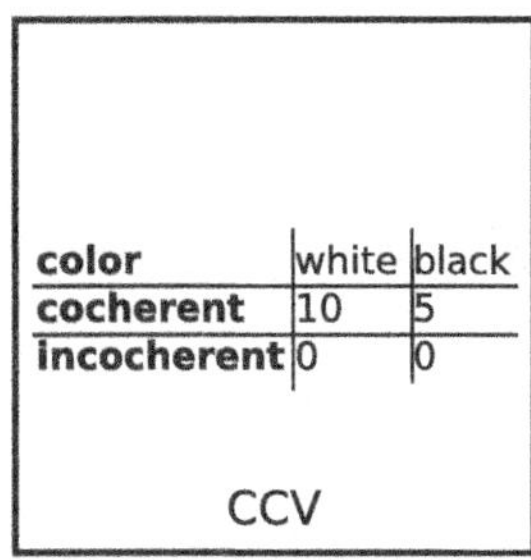

color	white	black
cocherent	10	5
incocherent	0	0

Fig. 3.17 Example results of the SURF and CCV algorithms. For the presented triangle, SURF detects three separate keypoints, and the CCV description is very vague

directly each to each. The presented descriptor allows omitting some comparison operations.

- Creating a descriptor that will be small and normalised. In the SURF algorithm, descriptors that characterise the surrounding neighbourhood of keypoints contain 64 floating-point values, whereas in the proposed method we reduce this amount of data to speed up the comparison and to minimise memory usage.

3.4.1 Method Description

We propose a method [22] that is a combination of local and global features and is focused on colour images to describe image patterns. It can also work with grayscale images as is with HOG and most of the keypoint-based methods but with worse results as we lose the colour information. The proposed descriptor refers slightly to the HOG and CCV algorithms (Sect. 2.6) but works differently.

3.4.1.1 Image Processing

In the proposed method, image features are extracted during the colour segmentation process which, in our case, is not preceded by any additional processing such as smoothing. Thus, the complete feature extraction is performed in a single pass, contrary to local keypoint-based methods that use multiple passes for size detection. An image is divided into regular fragments similarly to spatial pyramid-based algorithms [3, 16, 44]. The method performs segmentation and extraction of descriptors, which were collected from each area, into a single set.

During the segmentation stage, the method counts the number of segments and the number of segment's pixels in each colour group. The number of colours of

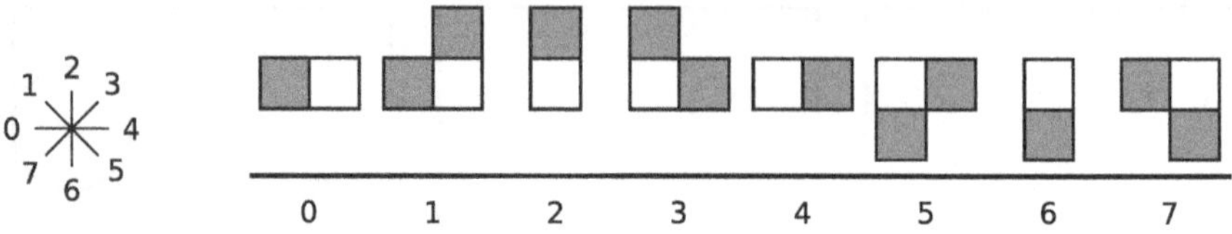

Fig. 3.18 Histogram of border directions

each segment is decreased to 64-colour space to reduce the combination of segments and memory usage. In this process, borders between segments are also counted as histograms where the position in the histogram corresponds to the direction of the edge created by the boundary (Fig. 3.18). Histograms of borders are counted for each colours combination; thus, they do not precisely represent the segments' shape but the distribution of the relationship between colours.

Thanks to colour space reduction to 64, the method during any image processing needs an array of 64x2 variables for image colour counting (number of segments, numbers of pixels) and also an array of 64x64x4 variables for border histograms counting (half of the eight-value histogram).

After image fragment processing, the method selects up to 10 most significant colours. For each of the selected colours, the method generates the main colour descriptor. After that, for each colour, we create sub-descriptors of colour relationships.

3.4.2 Color Descriptor

The first (primary) descriptor represents the colour that appears in an image fragment by describing colour domination and fragmentation. By this, it is possible to distinguish solid or sprayed colour patterns and their participation in the image. The descriptor contains three base values (Fig. 3.19). Some of them are normalised into the range 0–255 for efficient comparison and memory usage. The first is the colour number (C) for descriptors sorting or identification. Only descriptors of similar colour are compared. The second block consists of normalised values describing colour domination (D) and fragmentation (F). The domination is the ratio of the number of pixels of the same colour with respect to the number of all pixels in the sector. The fragmentation is the number of segments scaled compared to the number of colour pixels. Additional three values represent descriptor relation between sectors. Values of min and max y describe the range of vertical descriptor distribution over sectors. And the last value (sector count) describes a number of sectors where this descriptor occurred. Thus, the primary descriptor requires only 6 bytes of memory. Figure 3.20 presents descriptors for a set of simple example images. Each image contains only two colors; thus, we have only two descriptors. Colour 0 is black, and 63 is white. As we can see, it is possible to distinguish images which are more fragmented and which colour is dominant.

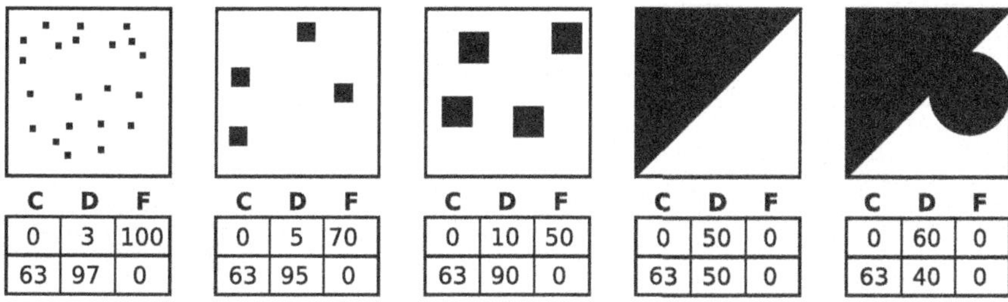

Fig. 3.19 Structure of the primary descriptor

Fig. 3.20 Examples of the primary descriptor

Fig. 3.21 Structure of the sub-descriptor

3.4.3 Colour Relationship Sub-descriptor

Using the primary descriptor, we are not able to compare and distinguish squared or rounded segments. To resolve this, a sub-descriptor was designed to describe the structure of colour relationships to other colours. The proposed sub-descriptor is closely related to the single main descriptor and indicates border-related colour. Figure 3.21 presents the descriptor structure. The first value of the descriptor is the aforementioned colour; the second is a normalised value of domination of this colour compared to other sub-descriptors. The last eight values constitute a normalised histogram of border edge directions.

Figure 3.22 presents examples of sub-descriptors. Tables under images contain related the main descriptor colour values (MC), sub-descriptor colours (C), relation dominations (D), histograms of relationship directions (H). As we see in this example, by using this additional description, it is possible to distinguish between the structure of patterns, such as, e.g. circles, squares or lines. Also, the gradient pattern can be described as a relation in a single similar direction between different segments.

3.4.4 Descriptor Comparison

Image comparing based on the proposed descriptors is more complex than in the case of the mentioned earlier local and global methods. Local feature-based methods

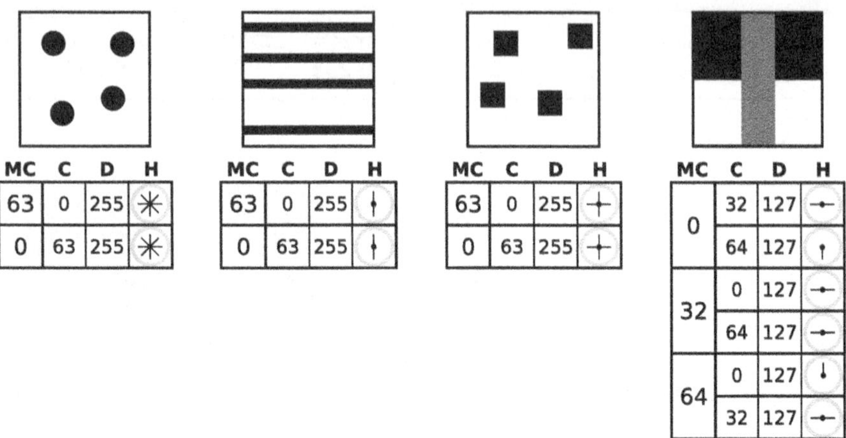

Fig. 3.22 Example of sub-descriptors

describe image fragments very precisely, thus, when some descriptors will be equal between images we could say that the image contains exactly similar content. Global methods such as CCV, generate a single descriptor per image, which describes the entire colour space; thus, it can be compared directly based on the distance between vectors. In our method, we divide it into smaller descriptors that describe only present colours. In this approach, we lose information of colours which are not present in the image. This could be problematic because the proposed method compares descriptors of similar colour and many images could be matched by a single common descriptor, even if other not common descriptors dominated the image. Because of this, our method checks the level of dissimilarity that is similar to a relative error. The comparison algorithm uses two counters of weighed descriptors number. The first one counts all descriptors. The weight of descriptors is ranked by their strength, and it is calculated by formula (3.38) where D is domination and SC is sector count.

$$C = D * SC \qquad (3.38)$$

Descriptors' similarity is checked by modulo distance matching with value thresholding, similarly to other feature descriptors. At first, the method checks the main descriptors, and if the distance thresholding passes then, the algorithm starts comparing linked sub-descriptors in a similar way. If the count of the sub-descriptors modulo distance passes the second thresholding, then, the entire descriptor is marked as similar.

3.4.5 Experimental Results

We performed experiments on images with various levels of similarity presented in Fig. 3.23 to compare the proposed method with the SURF algorithm (Sect. 2.12).

Fig. 3.23 Set of images used in the experiments in Sect. 3.4

Table 3.9 Results of image descriptors extraction from Fig. 3.23

The proposed method			SURF	
Descriptors	Sub-descriptors	Memory (KB)	Descriptors	Memory (KB)
31	87	1.031	134	33.500
39	102	1.225	153	38.250
33	79	0.965	69	17.250
70	299	3.330	156	39.000
37	117	1.359	150	37.500
26	41	0.553	42	10.500
45	81	1.055	72	18.000
23	38	0.506	29	7.250
8	6	0.105	0	0.000
3	2	0.037	0	0.000
18	28	0.379	0	0.000
7	6	0.100	11	2.750
142	794	8.586	9910	2477.500
153	707	7.801	4893	1223.250
147	618	6.896	3648	912.000
155	737	8.105	5806	1451.500
154	913	9.818	1077	269.250
156	875	9.459	760	190.000
155	961	10.293	1125	281.250
160	963	10.342	1008	252.000
158	1029	10.975	1276	319.000
159	1024	10.932	1255	313.750
151	975	10.406	1324	331.000
158	876	9.480	1011	252.750
160	858	9.316	1010	252.500
160	828	9.023	940	235.000
153	960	10.271	1334	333.500
96	353	4.010	39	9.750
156	1011	10.787	1760	440.000
152	897	9.650	2153	538.250
145	957	10.195	1945	486.250
152	957	10.236	1866	466.500
142	879	9.416	3819	954.750

The proposed method was implemented in C++ language with the Qt library. The SURF algorithm was also developed in C++, but it was based on the OpenSURF and OpenCV library. Both applications were single-threaded and were run on the same machine.

Table 3.9 presents the results of feature extraction from the test images presented in Fig. 3.23. As we can see, our method consumes much less memory than SURF because our descriptors are simpler and parameters are scaled to a single byte value versus the SURF descriptors of 64 float numbers.

In our method, the number of extracted descriptors increases with increasing number of colours and their combinations but not with the image size. In the SURF algorithm, the number of descriptors increases rapidly with the image size and complexity. In this experiment, the number of main descriptors and the sub-descriptors sum was about 46% of the SURF descriptors. Moreover, our method consumes about 1.6% memory compared to SURF.

The SURF algorithm perfectly locates and describes single characteristic points of images but achieves worse results on complicated patterns. Our method inversely reaches better results with complicated patterns, and single elements even can be omitted. It is because the method extracts descriptors from a determined sector in contrast to SURF that at first performs keypoint localisation.

3.4.6 Conclusions

After analysing the results of our experiments, we can claim that the new image descriptor is efficient in terms of memory usage and feature extraction and comparison speed versus, e.g. SURF. The new method describes images in a more detailed way than CCV but less than SURF, whereby it could be used to fast search for similar images without the necessity to contain exactly the same content. It could compare images by pattern content in an initial prefiltering process to speed up a more complex method. It could be used in a similar way to the HOG algorithm in spatial pyramid-based methods in content classification applications because of similar advantages such as merging descriptors from sectors to describe a larger area.

3.5 Fast Dictionary Matching

This section describes a method for searching for common sets of descriptors between collections of images. The presented method [24] operates on local interest keypoints, which are generated using the SURF algorithm. The use of a dictionary of descriptors allowed achieving good performance of the image retrieval. The method can be used to initially determine a set of similar pairs of keypoints between images. For this purpose, we use a certain level of tolerance between values of descriptors, as values of feature descriptors are almost never equal but similar between different images. After that, the method compares the structure of rotation and location of interest points in one image with the keypoint structure in other images. Thus, we were able to find similar areas in images and determine the level of similarity between them, even if the images contain different scenes.

Table 3.10 Differences between V_{sub} of two similar keypoint descriptors presented in Fig. 3.24

$V_{sub}x/y$	1	2	3	4
1	0.0000	0.0059	0.0031	−0.0047
2	−0.0098	0.0144	0.0349	0.0159
3	−0.0495	−0.0214	−0.0159	0.0079
4	−0.0770	−0.0062	−0.0120	−0.0173

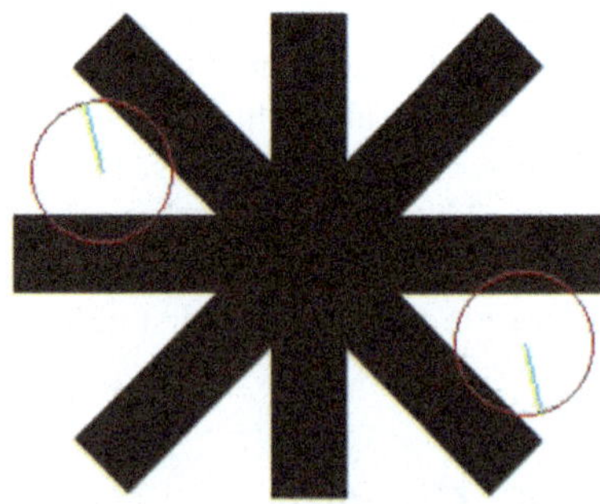

Fig. 3.24 An example of similar SURF keypoints with 0.47 value of difference between descriptor components

3.5.1   Description of the Problem

Usually, in order to compare local descriptor vectors, we have to apply some threshold considering their difference. Otherwise, it will be almost impossible to match keypoints from different images. For example, Table 3.10 presents a distribution of value differences of similar descriptors (Fig. 3.24) with the sum of absolute differences (SAD, L^1 norm, [13, 28]) equal to 0.4753. In this case, we consider the keypoints with SAD lower than 0.5 as similar. Presented keypoints and their orientations are identical for humans, but according to the values of descriptors, they are different. To compare two images, we have to compute two sets of keypoints. The number of keypoints depends on the size of images and the number of details. Often, for images larger than, e.g. a million pixels, the number of keypoints exceeds 1000. The easiest and the most common approach of comparison of keypoints between images is to compare each keypoint with the rest, but when we deal with a large number of keypoints, the number of needed computations is very high. For example, a thousand keypoints implicate 1 million comparisons. The keypoints should be ordered in some way to reduce the number of comparisons. Moreover, some of them should be omitted during the comparison process.

Another challenge when locating similar parts of images is the problem of keypoints being lost during image transformations. The cause of this problem is a different configuration of the same keypoints after the transformation. Usually, images representing the same content generate only a part of similar keypoints, another image can contain of course a different set of keypoints.

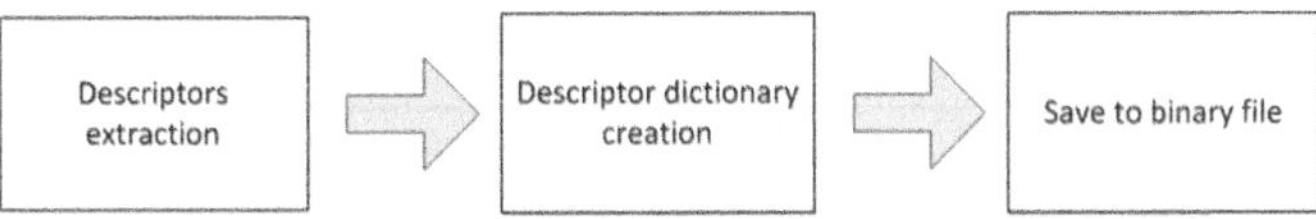

Fig. 3.25 Flowchart of the dictionary creation process

3.5.2 Method Description

For better performance, the proposed method uses a special, dictionary-based form of keypoint representation [7, 23]. Dictionary-based structures accelerate the comparison process by allowing to skip most of the keypoint combinations.

3.5.2.1 Dictionary Creation

Before matching images, the method detects keypoints and generates the dictionary structure for every single image (see Fig. 3.25).

The dictionary of keypoints is created from 64-element vectors which are local interest point descriptors of an image. The method puts separate elements of the descriptor in the dictionary beginning from the first element. The dictionary is built in a similar way to the B-tree [2], where the first element of the dictionary contains the list of first elements of descriptors.

The elements of descriptors which are similar and their values do not exceed estimated limits, are grouped and will be represented as a single element of the dictionary. An example of grouping is presented in Fig. 3.26 for the first element of descriptors with the number between 2 and 6. The rest of the descriptor elements, from which other elements are built, are derivatives of the first group. Thanks to grouping, we can decrease the number of similar, duplicated elements of descriptors. Thanks to the presented approach, building index of descriptors is also faster, especially when we deal with a very large number of descriptors. The rest of the keypoint descriptor data, such as position, size or orientation are contained in the last part of the word associated with the descriptor. The last step of the process of creation of the dictionary is a conversion of data to a binary file as it is sufficient to generate the dictionary only once.

3.5.3 Comparison Between Descriptors and Dictionary

Every image from the analysed set has its own descriptor dictionary stored in the form of a binary file (see Sect. 3.5.2). Now, let us assume that we have a new query

<table>
<thead>
<tr><th rowspan="2">Descriptor Number</th><th colspan="5">Number of descriptor element</th></tr>
<tr><th>1</th><th>2</th><th>3</th><th>4</th><th>5</th></tr>
</thead>
<tbody>
<tr><td>1</td><td>-0.1021</td><td>0.2031</td><td>0.1229</td><td>-0.0551</td><td>0.2168</td></tr>
<tr><td>2</td><td rowspan="5">0.0451</td><td rowspan="3">0.1846</td><td rowspan="3">0.2229</td><td>-0.0551</td><td>-0.0720</td></tr>
<tr><td>3</td><td rowspan="2">0.1779</td><td>-0.0068</td></tr>
<tr><td>4</td><td>0.2168</td></tr>
<tr><td>5</td><td rowspan="2">0.4722</td><td>-0.1021</td><td>-0.0067</td><td>0.0205</td></tr>
<tr><td>6</td><td>-0.0551</td><td>0.2168</td><td>0.1780</td></tr>
<tr><td>7</td><td rowspan="2">0.2687</td><td>0.1846</td><td>-0.0067</td><td>0.0205</td><td>-0.0068</td></tr>
<tr><td>8</td><td>0.5151</td><td>0.2168</td><td>0.1779</td><td>0.2168</td></tr>
<tr><td>9</td><td rowspan="2">0.4552</td><td rowspan="2">0.1847</td><td>-0.0068</td><td>-0.0551</td><td>-0.0721</td></tr>
<tr><td>10</td><td>0.0205</td><td>-0.0068</td><td>0.2168</td></tr>
</tbody>
</table>

Fig. 3.26 A part of the descriptor dictionary example

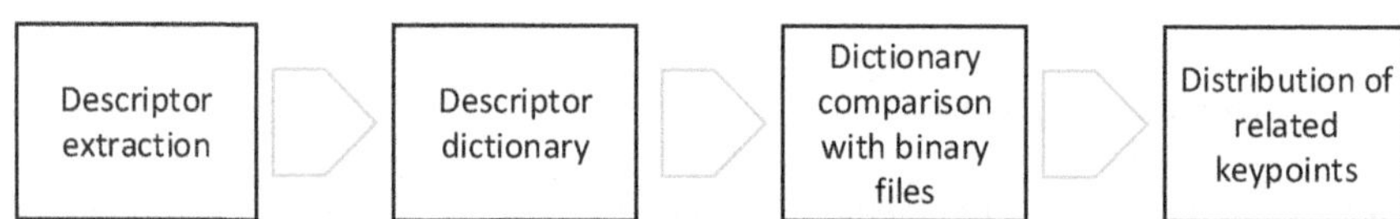

Fig. 3.27 Flowchart of the image retrieval in the set of images

image and we want to find similar images in a large collection of images. The first step is to create a dictionary of its feature descriptors and store it in a binary file. Figure 3.27 presents a flowchart of such image retrieval. The next step is a comparison of the query image dictionary with the dictionaries from the binary files. Descriptors values are similar if their sum of absolute differences (SAD) is less than the threshold. Comparison of two dictionaries is presented in Fig. 3.28, where the dark background represents a common part.

3.5.4 Matching Sets of Keypoints

The dictionary comparing process returns a set of pairs of similar keypoints. The next step is to examine keypoint distribution between images. Each pair will be excluded if their distribution in relation to the rest of the pairs indicates a wrong connection. Figure 3.29 describes an example of keypoint distribution between two images. Each point has its own counterpart in the second set. The method compares the direction and the distance between keypoints from the same set. For example, angles β_{12} and α_{12} have the same value as β_{12}' and α_{12}' from the second set. Distances d_{12} and d_{12}' are also similar. Thus, in this case we can assume, that points P_1 and P_2 are related. Otherwise, we mark keypoints as not related, e.g. P_4 and P_4'.

Number of descriptor element

Descriptor Number	1	2	3	4	5
1	-0.1021	0.2031	0.1229	-0.0551	0.2168
2	0.0451			-0.0551	-0.072
3		0.1846	0.2229	0.1779	-0.0068
4					0.2168
5		0.4722	-0.1021	-0.0067	0.0205
6			-0.0551	0.2168	0.1780
7	0.2687	0.5151	0.2168	0.1779	0.2168
8		0.1846	-0.0067	0.0205	-0.0068
9	0.4552	0.1846	-0.0068	-0.0551	-0.0720
10			0.0205	-0.0068	0.2168

Dictionary form binary file

Number of descriptor element

Descriptor Number	1	2	3	4	5
1	0.042	0.5111	-0.1021	-0.0067	0.0205
2			-0.0551	0.2168	0.1780
3	0.1187	0.1846	-0.0067	0.0205	-0.0068
4		0.5151	0.2168	0.1779	0.2168
5	0.2687	0.1846	-0.0205	-0.0068	0.2168
6			-0.0068	0.03	-0.0068

Fig. 3.28 Example of a part of two compared dictionaries

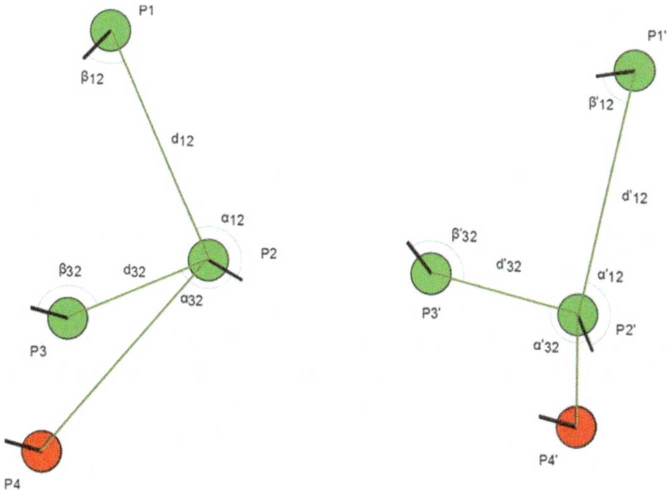

Fig. 3.29 Example of keypoint pair checking by mapping them between images

Table 3.11 Results of comparisons between image dictionaries

Images	No. of points	Matched	Comparisons	Combinations	Performance (%)
01	367	225	19522	3733124	0.52
02	257	26	3691	2614204	0.14
03	727	103	15373	7395044	0.21
04	80	101	1747	813760	0.21
05	408	112	10773	4150176	0.26
06	24	22	413	244128	0.17
07	729	0	0	7415388	0.00
08	414	20	7676	4211208	0.18
09	845	20	7674	8595340	0.09
10	359	128	5137	3651748	0.14
11	318	128	5107	3234696	0.16
12	213	44	3815	2166636	0.18
13	891	52	13049	9063252	0.14
14	785	61	19567	7985020	0.25
15	435	162	10068	4424820	0.23
16	295	95	10575	3000740	0.35
17	489	154	10408	4974108	0.21
18	650	116	14754	6611800	0.22
19	417	186	13569	4241724	0.32
20	464	104	13479	4719808	0.29
21	1005	5	134	10222860	0.00

3.5.5   *Experimental Results*

In this section, we show some examples of the proposed method for content-based image retrieval on the test images presented in Fig. 3.30. For better presentation, we chose images, which are only slightly different. Table 3.11 contains test results of comparisons between each image with all other from Fig. 3.30. "No. of points" column is the number of descriptors extracted from the image. "Matched" column is the number of related descriptors between the current one and all other images. "Comparisons" is the number of descriptors compared by using the dictionary. "Combination" is the number of all possible combinations of descriptors between images. As we can see, the number of comparisons in the proposed method is significantly smaller than the number of all combinations. In our tests, the number of compared descriptors is only 0.18% of all possible combinations.

Figure 3.31 presents the results of search for common set of keypoints from image number 17. The largest image is our query image. The others are found similar

Fig. 3.30 Images used in the experiments in Sect. 3.5

Fig. 3.31   Example of the detected groups of descriptors between images

Table 3.12   Results of the comparison between the images from Fig. 3.31

No. image (query)	No. of keypoints	No. image (compared)	No. of keypoints	Math	Comparisons
17	489	4	80	31	290
17	489	15	435	17	1491
17	489	18	650	20	2197
17	489	19	417	57	1708
17	489	20	464	23	1723

images. Related points are marked on each image. Larger points are the centers of keypoints that describe common area.

Table 3.12 presents detailed values from the comparison procedure between the images from Fig. 3.31. Only part of keypoints were connected, but this number allows selecting a common part of both images. In the presented case, a single image was incorrectly marked as related to the query image. It was caused by a similarity between descriptors and their orientation.

3.5.6   Conclusions

Analysing the results of our experiments, we can say that the creation of the dictionary allows to significantly decrease the number of operations, which have to be done in the process of image comparison. In our case, the number of operation has been reduced to 18% of all operations. The approach obtains better results in the case of larger sets of images. Images related to the query image can be found much faster in comparison to the standard all-to-all, brute-force matching. Moreover, saving the dictionary in a binary file allows for more efficient image multiple comparisons and reuse of data.

3.6 Summary and Discussion

There are many ways to compare features, from vector distance measures to fuzzy set-related [14, 15]. In some cases, it is required to classify images by their content. To this end, many methods were developed that could learn a combination of image features specific for a visual class. In the case of image classification, usually visual feature extraction methods are combined with machine learning, e.g. with support vector machines [36] or artificial neural networks. Classifiers in most cases at first need to be trained by a set of prepared data of known classes. Sometimes images are divided into regular sectors, and within them, descriptors are generated, and classifiers are trained. Local feature-based algorithms can be also used with a spatial localisation of keypoints [26]. Global feature algorithms are far easier applicable to classification because of the constant number of feature data per image.

This chapter presented some techniques accelerating image matching process. The first method generated fuzzy rules from the most representative local features for each class by the AdaBoost algorithm. Compared to the bag of features algorithm, the presented method was more accurate and faster. Then, a novel method for automatic image description based on local image features was proposed. As the resulting descriptor is concise, image and object comparison during retrieval is relatively fast. Such a descriptor is relatively fast to compute and immune to scale change or rotation. The method provides the angle-keypoint histogram as an object descriptor. This chapter also presented a method to speed up image content similarity search with the SURF algorithm in large sets of images based on descriptor dual-hashing. The method stores SURF extracted keypoints in an ordered way that significantly reduces the number of unnecessary comparisons between sets of image keypoints during matching. The data structure is optimised to be stored in a file and to search without pre-loading to avoid image feature extraction in each search process. Moreover, the chapter described a new image descriptor based on colour spatial distribution for image similarity comparison. It is similar to methods based on HOG and spatial pyramid but in contrast to them operates on colours and colour directions instead of oriented gradients. The presented method assumes using two types of descriptors. The first one is used to describe segments of similar colour and the second sub-descriptor describes connections between different adjacent segments. Thus, the image parts can be described in a more detailed way as is in the case of the histogram of oriented gradients (HOG) algorithm but more general as is in the case of keypoint-based methods such as SURF or SIFT. Moreover, in comparison to the keypoint-based methods, the proposed descriptor is less memory demanding and needs only a single step of image data processing. Descriptor comparing is more complicated but allows for descriptor ordering and for avoiding some unnecessary comparison operations.

The last method presented in this chapter was dictionary searching for common sets of local descriptors between collections of images. The use of a dictionary of descriptors allowed achieving good performance of the content-based image retrieval. The method can be used to initially determine a set of similar pairs of keypoints between images. It is possible to find similar areas in images and determine the level of similarity between them, even when images contain different scenes. The performed experiments showed that the introduced set of algorithms is efficient in terms of speed and accuracy. The approaches can be applied to various computer vision systems. Although slightly less accurate, the proposed algorithms are much faster than the solutions based on convolutional neural networks.

References

1. Bay, H., Ess, A., Tuytelaars, T., Van Gool, L.: Speeded-up robust features (surf). Comput. Vis. Image Underst. **110**(3), 346–359 (2008)
2. Bayer, R., McCreight, E.M.: Organization and maintenance of large ordered indexes. Acta Informatica **1**(3), 173–189 (1972). https://doi.org/10.1007/BF00288683
3. Bosch, A., Zisserman, A., Munoz, X.: Representing shape with a spatial pyramid kernel. In: Proceedings of the 6th ACM International Conference on Image and Video Retrieval, pp. 401–408. ACM (2007)
4. Bradski, G.: The opencv library. Dr. Dobbs J. **25**(11), 120–126 (2000)
5. Buckland, M., Gey, F.: The relationship between recall and precision. J. Am. Soc. Inf. Sci. **45**(1), 12 (1994)
6. Datar, M., Immorlica, N., Indyk, P., Mirrokni, V.S.: Locality-sensitive hashing scheme based on p-stable distributions. In: Proceedings of the Twentieth Annual Symposium on Computational Geometry, SCG 2004, pp. 253–262. ACM, New York, NY, USA (2004)
7. Edelkamp, S., Schroedl, S.: Heuristic Search: Theory and Applications. Elsevier (2011)
8. Everingham, M., Van Gool, L., Williams, C.K.I., Winn, J., Zisserman, A.: The pascal visual object classes (voc) challenge. Int. J. Comput. Vis. **88**(2), 303–338 (2010)
9. Grauman, K., Darrell, T.: Efficient image matching with distributions of local invariant features. In: Computer Vision and Pattern Recognition, 2005. CVPR 2005. IEEE Computer Society Conference on, vol. 2, pp. 627–634 vol. 2 (2005). https://doi.org/10.1109/CVPR.2005.138
10. Grycuk, R., Gabryel, M., Scherer, M., Voloshynovskiy, S.: Image descriptor based on edge detection and crawler algorithm. In: International Conference on Artificial Intelligence and Soft Computing, pp. 647–659. Springer International Publishing (2016)
11. Grycuk, R., Gabryel, M., Scherer, R., Voloshynovskiy, S.: Multi-layer architecture for storing visual data based on WCF and microsoft sql server database. In: Artificial Intelligence and Soft Computing, Lecture Notes in Computer Science, vol. 9119, pp. 715–726. Springer International Publishing (2015)
12. Grycuk, R., Gabryel, M., Scherer, R., Voloshynovskiy, S.: Multi-layer architecture for storing visual data based on WCF and microsoft sql server database. In: International Conference on Artificial Intelligence and Soft Computing, pp. 715–726. Springer International Publishing (2015)
13. Hamzah, R.A., Rahim, R.A., Noh, Z.M.: Sum of absolute differences algorithm in stereo correspondence problem for stereo matching in computer vision application. In: 2010 3rd International Conference on Computer Science and Information Technology, vol. 1, pp. 652–657 (2010). https://doi.org/10.1109/ICCSIT.2010.5565062

14. Korytkowski, M.: Novel visual information indexing in relational databases. Integr. Comput. Aided Eng. **24**(2), 119–128 (2017)
15. Korytkowski, M., Rutkowski, L., Scherer, R.: Fast image classification by boosting fuzzy classifiers. Information Sciences **327**, 175–182 (2016). https://doi.org/10.1016/j.ins.2015.08.030. URL http://www.sciencedirect.com/science/article/pii/S0020025515006180
16. Lazebnik, S., Schmid, C., Ponce, J.: Beyond bags of features: Spatial pyramid matching for recognizing natural scene categories. In: 2006 IEEE Computer Society Conference on Computer Vision and Pattern Recognition, vol. 2, pp. 2169–2178. IEEE (2006)
17. Lowe, D.G.: Distinctive image features from scale-invariant keypoints. Int. J. Comput. Vis. **60**(2), 91–110 (2004)
18. Matas, J., Chum, O., Urban, M., Pajdla, T.: Robust wide-baseline stereo from maximally stable extremal regions. Image Vis. Comput. **22**(10), 761–767 (2004). British Machine Vision Computing 2002
19. Meskaldji, K., Boucherkha, S., Chikhi, S.: Color quantization and its impact on color histogram based image retrieval accuracy. In: Networked Digital Technologies, 2009. NDT 2009. First International Conference on, pp. 515–517 (2009). https://doi.org/10.1109/NDT.2009.5272135
20. Mikolajczyk, K., Schmid, C.: Scale and affine invariant interest point detectors. Int. J. Comput. Vis. **60**(1), 63–86 (2004)
21. Najgebauer, P., Grycuk, R., Scherer, R.: Fast two-level image indexing based on local interest points. In: 2018 23rd International Conference on Methods Models in Automation Robotics (MMAR), pp. 613–617 (2018). https://doi.org/10.1109/MMAR.2018.8485831
22. Najgebauer, P., Korytkowski, M., Barranco, C.D., Scherer, R.: Artificial Intelligence and Soft Computing: 15th International Conference, ICAISC 2016, Zakopane, Poland, June 12–16, 2016, Proceedings, Part II, chap. Novel Image Descriptor Based on Color Spatial Distribution, pp. 712–722. Springer International Publishing, Cham (2016)
23. Najgebauer, P., Nowak, T., Romanowski, J., Gabryel, M., Korytkowski, M., Scherer, R.: Content-based image retrieval by dictionary of local feature descriptors. In: 2014 International Joint Conference on Neural Networks, IJCNN 2014, Beijing, China, July 6–11, 2014, pp. 512–517 (2014)
24. Najgebauer, P., Rygal, J., Nowak, T., Romanowski, J., Rutkowski, L., Voloshynovskiy, S., Scherer, R.: Fast dictionary matching for content-based image retrieval. In: Rutkowski, L., Korytkowski, M., Scherer, R., Tadeusiewicz, R., Zadeh, L.A., Zurada, J.M. (eds.) Artificial Intelligence and Soft Computing, Lecture Notes in Computer Science, vol. 9119, pp. 747–756. Springer International Publishing (2015)
25. Nister, D., Stewenius, H.: Scalable recognition with a vocabulary tree. In: Proceedings of the 2006 IEEE Computer Society Conference on Computer Vision and Pattern Recognition - Volume 2, CVPR 2006, pp. 2161–2168. IEEE Computer Society, Washington, DC, USA (2006)
26. Nowak, T., Najgebauer, P., Romanowski, J., Gabryel, M., Korytkowski, M., Scherer, R., Kostadinov, D.: Spatial keypoint representation for visual object retrieval. In: Artificial Intelligence and Soft Computing, Lecture Notes in Computer Science, vol. 8468, pp. 639–650. Springer International Publishing (2014)
27. Philbin, J., Chum, O., Isard, M., Sivic, J., Zisserman, A.: Object retrieval with large vocabularies and fast spatial matching. In: Computer Vision and Pattern Recognition, 2007. CVPR 2007. IEEE Conference on, pp. 1–8 (2007)
28. Richardson, I.E.: H. 264 and MPEG-4 Video Compression: Video Coding for Next-generation Multimedia. Wiley (2004)
29. Rublee, E., Rabaud, V., Konolige, K., Bradski, G.: Orb: An efficient alternative to sift or surf. In: 2011 IEEE International Conference on Computer Vision (ICCV), pp. 2564–2571 (2011). https://doi.org/10.1109/ICCV.2011.6126544
30. Rutkowski, L.: Flexible Neuro-Fuzzy Systems. Kluwer Academic Publishers (2004)
31. Rutkowski, L.: Computational Intelligence Methods and Techniques. Springer, Berlin, Heidelberg (2008)

32. Schapire, R.E.: A brief introduction to boosting. In: Proceedings of the 16th International Joint Conference on Artificial Intelligence - Volume 2, IJCAI 1999, pp. 1401–1406. Morgan Kaufmann Publishers Inc., San Francisco, CA, USA (1999)
33. Scherer, R.: Designing boosting ensemble of relational fuzzy systems. Int. J. Neural Syst. **20**(5), 381388 (2010). http://www.worldscinet.com/ijns/20/2005/S0129065710002528.html
34. Scherer, R.: Multiple Fuzzy Classification Systems. Springer Publishing Company, Incorporated (2014)
35. Sivic, J., Zisserman, A.: Video google: a text retrieval approach to object matching in videos. In: Proceedings of the Ninth IEEE International Conference on Computer Vision, 2003. vol. 2, pp. 1470–1477 (2003)
36. Sopyla, K., Drozda, P., Górecki, P.: Svm with cuda accelerated kernels for big sparse problems. In: ICAISC (1), Lecture Notes in Computer Science, vol. 7267, pp. 439–447. Springer (2012)
37. Tao, D.: The corel database for content based image retrieval (2009)
38. Tao, D., Li, X., Maybank, S.J.: Negative samples analysis in relevance feedback. IEEE Trans. Knowl. Data Eng. **19**(4), 568–580 (2007)
39. Tao, D., Tang, X., Li, X., Wu, X.: Asymmetric bagging and random subspace for support vector machines-based relevance feedback in image retrieval. IEEE Trans. Pattern Anal. Mach. Intell. **28**(7), 1088–1099 (2006)
40. Tieu, K., Viola, P.: Boosting image retrieval. Int. J. Comput. Vis. **56**(1–2), 17–36 (2004)
41. Ting, K.M.: Precision and recall. In: Encyclopedia of Machine Learning, pp. 781–781. Springer (2011)
42. Viola, P., Jones, M.: Rapid object detection using a boosted cascade of simple features. In: Proceedings of the 2001 IEEE Computer Society Conference on Computer Vision and Pattern Recognition, 2001. CVPR 2001, vol. 1, pp. I–511–I–518 (2001)
43. Voloshynovskiy, S., Diephuis, M., Kostadinov, D., Farhadzadeh, F., Holotyak, T.: On accuracy, robustness, and security of bag-of-word search systems. In: IS&T/SPIE Electronic Imaging, pp. 902, 807–902,807. International Society for Optics and Photonics (2014)
44. Yang, J., Yu, K., Gong, Y., Huang, T.: Linear spatial pyramid matching using sparse coding for image classification. In: IEEE Conference on Computer Vision and Pattern Recognition, 2009. CVPR 2009, pp. 1794–1801. IEEE (2009)
45. Zhang, J., Marszalek, M., Lazebnik, S., Schmid, C.: Local features and kernels for classification of texture and object categories: a comprehensive study. In: Conference on Computer Vision and Pattern Recognition Workshop, 2006. CVPRW 2006, pp. 13–13 (2006). https://doi.org/10.1109/CVPRW.2006.121
46. Zhang, W., Yu, B., Zelinsky, G., Samaras, D.: Object class recognition using multiple layer boosting with heterogeneous features. In: IEEE Computer Society Conference on Computer Vision and Pattern Recognition, 2005. CVPR 2005, vol. 2, pp. 323–330 vol. 2 (2005). https://doi.org/10.1109/CVPR.2005.251

Chapter 4
Novel Methods for Image Description

This chapter presents new methods for edge detection and description. Standard edge detection algorithms confronted with the human perception of reality are rather primitive because they are based only on the information stored in the form of pixels. Humans can see elements of the images that do not exist in them. These mechanisms allow humans to extract and track objects partially obscured. These rules are described in the form of many complementary Gestalt principles [4, 15] which allow us to understand how much the real picture is different from the one we perceive. Furthermore, this chapter proposes a novel method for describing continuous edges as vectors.

4.1 Algorithm for Discontinuous Edge Description

In this section, we present a method of predictive reconstructing connections between parts of object outlines in images [12]. The method was developed mainly to analyse microscopic medical images but applies to other types of images. Examined objects in such images are highly transparent; moreover, close objects can overlap each other. Thus, segmentation and separation of such objects can be difficult. Another frequently occurring problem is partial blur due to high image magnification. Large focal length of a microscope dramatically narrows the range of the sharp image (depth of field).

The method is based on edge detection to extract object contours and represent them in a vector form. The logic behind the presented method refers to the Gestalt Laws describing human perception. The method, according to the law of good continuation and the principle of similarity, evaluates the neighbourhood the interrupted contour path and then tries to determine the most appropriate connection with the other parts of the contour. To assess the similarity of contour parts, the method

© Springer Nature Switzerland AG 2020

R. Scherer, *Computer Vision Methods for Fast Image Classification and Retrieval*, Studies in Computational Intelligence 821,
https://doi.org/10.1007/978-3-030-12195-2_4

examines the orientation of the line determined by the gradient of the edge, the characteristics of the edge cross section and the direction of its current course.

In order to reduce the amount of data and accelerate the method, fragments of the detected outlines are represented by vectors in the form of a graph. Thus, the method has faster access to the information on the course of the edges than in the case of bitmap-based representations. The method presented in this section refers mainly to the Gestalt laws' principles of similarity and good continuation, aiming to link together fragments of a line in one of the object's outline. In this way, it is possible to extract obscured objects that even cannot be done by image segmentation.

Edge detection and image segmentation are used in this section in microscopic image analysis. The analyzed images come mostly from laboratory examination of biological samples viewed under a microscope. Microscopic examination is the diagnosis or evaluation of samples. Usually we deal with histopathological examination of tissue samples during cancer research [2, 5, 21]. They are designed to assess a sample for detection of cancer and type of malignancy in a segment of tissue. In this case, tissue and sought-after object structures are relatively uniform; thus the best to apply are methods based on segmentation or detection of consistent areas. Another area of application is the parasitological image analysis [3, 18–20, 24]. In this case, the image is scanned to search for worm eggs or adult forms, and it is a part of the diagnosis of parasitic diseases. Comparing to medical histopathological diagnosis, the samples are examined under less magnification, but certain three-dimensionality is added. In the case of histopathology, samples are cut flat whereas in this case, there are several layers, and individual objects have their height. With this type of images, contours of the searched objects are more important. The shape and structure of the parasite eggs walls are the basis for their classification.

For the parasite egg shape detection purposes, we could also apply methods such as active shape model [9, 17] because the most of the eggs have a similar shape. However, in the case of parasite egg detection, the objects can be distributed in random numbers, rotation and position that requires to perform a large number of checks. In order to solve this problem, our method tries to group or omit some content to reduce the number of combinations.

As aforementioned, in some cases the examined objects do not have a complete outline, e.g. can be partially obscured. The human mind tries to unconsciously trace the continuation of their elements that seem to be lost. Most often it relies on the idea of continuing the current course of the image edge (Gestalt rule of good continuation). It is a form of inertia of vision, i.e. if the eyesight loses the contact with the tracked object, it tries to continue to follow it. When the end of the partially obscured object is reached, humans try to invent a continuation (similarity rule). In the absence of continuation, the mind suggests that the object has a continuing fragment and is finally connected to the other end of the object (closure rule).

In the case of segmentation of the masked object edges, the method which follows along the line of the local maximal edge detector responses (Fig. 4.1b) is unable to separate them (Fig. 4.1a) or connect them. The effect may even be counterproductive, as two separate objects can be detected. This problem is encountered, e.g. in some

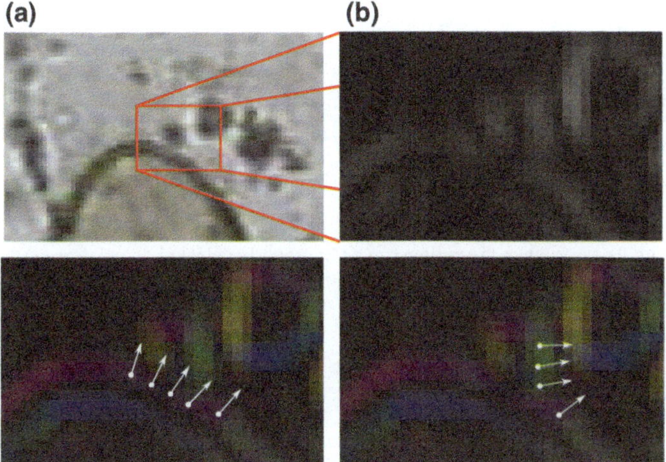

Fig. 4.1 An example of an improper edge segmentation. **a** Presents desirable but weaker contour. **b** Presents incorrectly detected fragment of the contour by the local maxima-based edge detector. White arrow represents edge detector result values

parasitical images, where the objects can touch or can overlap each other. Figure 4.2 presents an example of wrong segmentation, where, as we can see, the outline should be continued on the contour line of the egg. However, it deviates and follows the adjacent object (Fig. 4.2b). It is desired that during the segmentation the outline is interrupted at this point and linked to the suitable further fragment of the outline (Fig. 4.2c)). There are relatively more such cases, and each has its characteristic, as presented in Fig. 4.3. The methods developed for this purpose are most often based on binding edges on the basis of their distance and orientation [16, 22]. However, in some cases, the method finds several lines located close to the continuation in accordance with the orientation (Fig. 4.4). Thus, there is a problem with their proper

Fig. 4.2 An example of improperly detected object outline. **b** Presents incorrect direction of outline segmentation caused by the adjacent object. **c** Presents expected results by outline slice at the point of objects contact and contour direction change

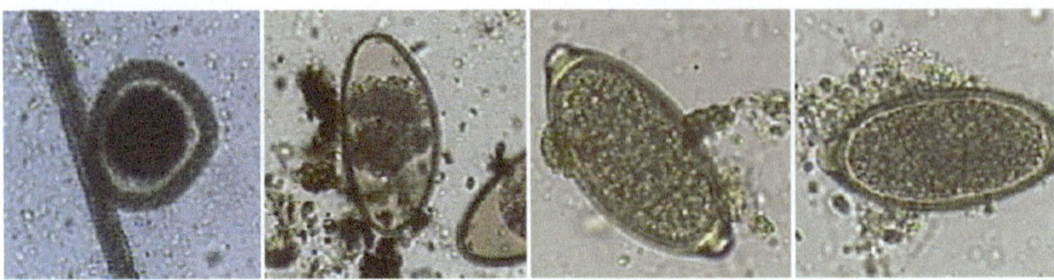

Fig. 4.3 Example cases when the outline can be improperly detected. In most cases, the outline is disturbed by adjacent or overlapped random pollution contained in microscopic samples

(a)   **(b)**   **(c)**

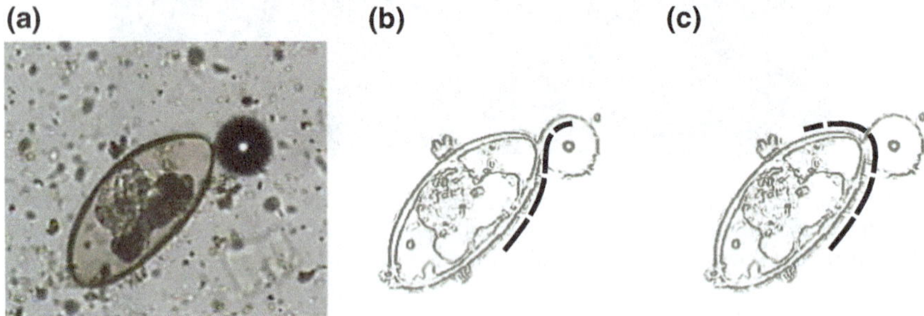

Fig. 4.4 An example of improperly detected outline continuation by connecting edge fragments. **b** Presents incorrect connection as edges fragments create a straighter line. If we consider a bigger image of all fragments then **c** presents a better continuation

connection. In this case, combining by the orientation of the lines may not give satisfactory results (Fig. 4.4b). Part of the so-created connections, when verified by the man, can be assessed as improper. This case is often seen in coproscopy images, where, as previously mentioned, samples contain impurities that can hinder the determination of the correct object outlines.

4.1.1 Proposed Approach

Providing better results for determining the outline of objects, we propose to extend the outline completion based on the orientation of edges with additional properties. Relying on the Gestalt laws, we propose to add an additional edge descriptor (the law of similarity) that allows specifying whether the combined fragments of the outline are indeed similar to each other, and this parameter determines the direction in which the edge should be further followed.

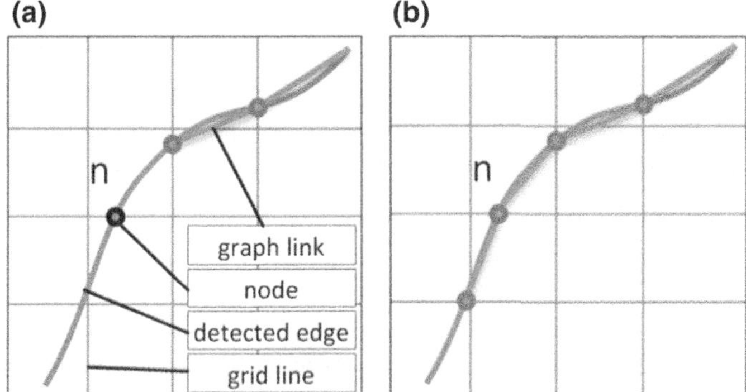

Fig. 4.5 Graph representation of edges. Each detected intersection of the grid lines with the detected edge is represented by a graph node. Graph links are determined by edge segmentation between nodes

4.1.1.1 Edge Graph Representation

In order to effectively operate on the edges of the image in our solution, it was decided to represent the detected edges as a graph, which creation method was more precisely described in our previous work [13]. The graph is built along the auxiliary grid and has a regular node layout (Figs. 4.5 and 4.9c), with a similar distance between nodes. An advantage of the graph method is the reduction of data amount to be considered in the next steps of the method. For example, our method represents a straight line of 200 pixels length by 20 nodes in the case of using 10-pixel grid size. Another advantage of the proposed solution in parasitology imaging is omitting pollution of smaller size than the grid size. This allows tracing the graph nodes mapping the edges and determine the parameters of the curvature of the edges. In addition, each node of the graph has its orientation corresponding to the direction of the edge (Fig. 4.3). The orientation vector can designate the edge descriptor to allow to compare it with other edges, even if they have a different orientation.

4.1.1.2 Curvature Coefficients

The curvature coefficients are the first parameters which allow specifying which parts of the outline better fit to each other. They map in a sense, the current course of the edge. In the proposed solution a vector of four coefficients determined according to the formula (4.1) as the average values of the angles formed between the nodes of the graph and their orientation. Thanks to them lines can be distinguished based on their curvature and resultant undulations.

$$cc = \left[\sum \alpha_i - \alpha_{i-1}, \ \sum |\alpha_i - \alpha_{i-1}|, \right.$$
$$\left. \sum \beta_i - \beta_{i-1}, \ \sum |\beta_i - \beta_{i-1}|\right] \tag{4.1}$$

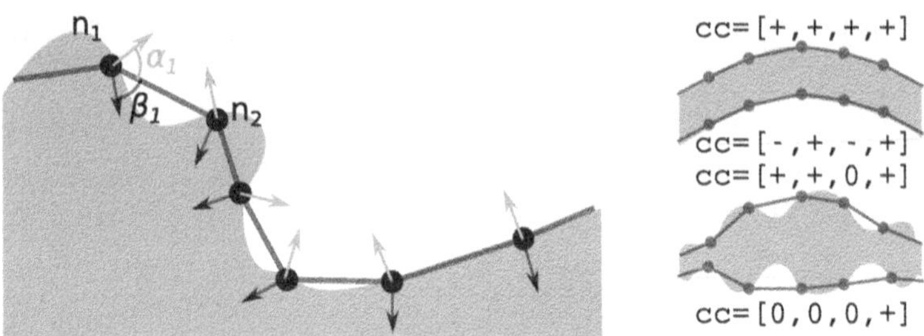

Fig. 4.6 The way of determining angles needed to compute curvature coefficients. The arrows represent the normal vector, respectively, to image edges (grey arrows) and graph links (black arrows) to previous graph nodes

In the proposed graph edge representation, each node is described by the coordinates and the normal vector of the edge in its position. This allows determining the angle values to calculate the curvature coefficients most effectively by a combination of scalar products of vectors. Figure 4.6 shows a diagram for determining the scalar products of angles. Directional vectors between adjacent nodes are reference vectors. The product of the angle α is determined relative to the reference vectors and the normal edge vector. The product of β is determined from the normal reference vector and the reference vector of the previous pair of nodes. Finally, the coefficients should be normalised to compare paths with different lengths effectively. After the normalisation, coefficient sums of the products take the values in the range from -1 to 1, and the absolute value of the sum of the products of from 0 to 2. The comparison can take place by determining the distance between curvature coefficients vectors. In the case of parasitological images, we most often deal with one-sided bent paths in relation to the walls of the worm eggs. However, in the event of pollution, the outlines can have arbitrary shapes. With the curvature coefficients, we can join the edges only by their similarity, but for the correctness of joining the edge pieces, we must first check their respective positions so that they form the whole object. In this case, the extreme nodes of the fragments are examined. According to Fig. 4.7, the edge position similarity is determined from the normal vectors of the extreme nodes and the direction vector between them. We determine the scalar product of the vectors, according to formula (4.2). All values are in the range of 0 to 2, where values close to 0 indicate a better position relative to each other. The first value we get tells us about the degree of parallelism, where 0 means that the pieces are in line or symmetric, while the value increases when the lines begin to have different directions. The second value specifies the mutual orientation of the line in the range where 0 determines the consistency of the node's direction, and two means that the edge nodes are opposite, for example, for edge fragments on the other side of the object.

$$oc = \left[|\cos \alpha_1 - \cos \alpha_2|,\ 1 - \cos \beta \right] \qquad (4.2)$$

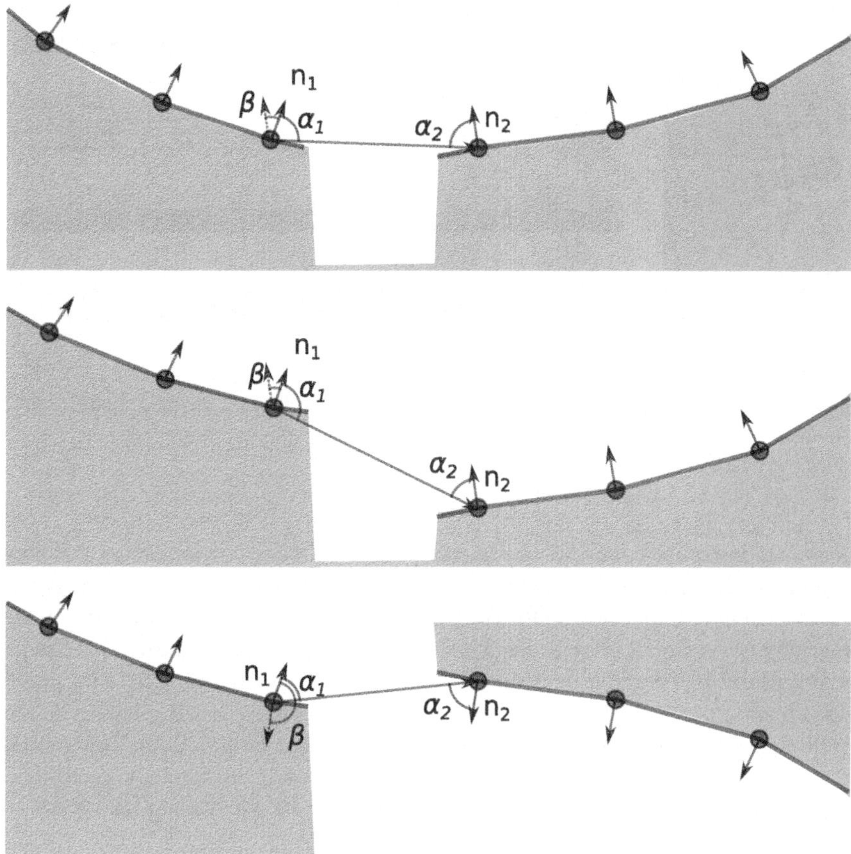

Fig. 4.7 Initial checking of line fragments orientation. **a** Presents a configuration of fragments that can create a proper and continuous line. **b** Presents fragments that become parallel and **c** presents opposing fragments

4.1.1.3   Edge Intersection Descriptor

Comparing the direction and the curvature of the edge cannot avoid the situation in which the edge may be incorrectly connected. Concerning the parasitological images, such cases may occur for, e.g. edges located close to the edges partitioning the inner layer of the parasite egg wall, or located close to the other eggs or contamination. When evaluating such a connection, we can easily see that the elements do not fit together. The edge environment itself strongly indicates their dissimilarity. For this reason, an additional element that allows achieving a better connection of edges is to add a simple descriptor for elements around the edges. Figure 4.8 shows the characteristics of the cross-section of the walls of several various parasites, allowing to differentiate the edges. The descriptor must be characterized by taking into account

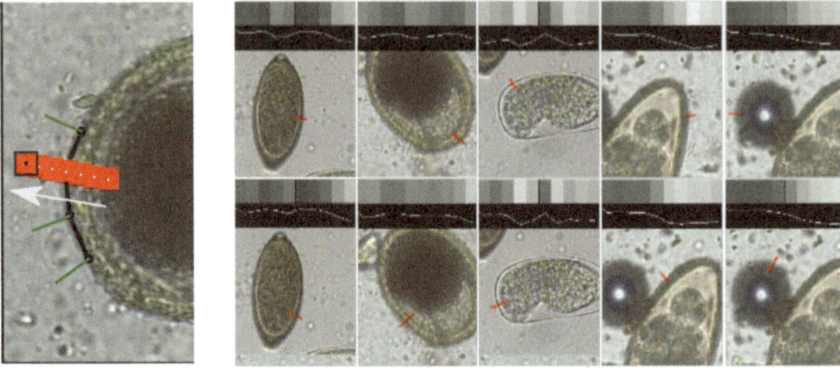

Fig. 4.8 The direction and method of edge descriptor determination with an example of edge descriptors for various parasite egg cells. The white arrow is a direction from the first descriptor value to the last that is similar to the node normal vector

the orientation of the edges, so we can easily compare them with the others as if they were in the same position and orientation. Figure 4.8 shows how the edge descriptor is evaluated in accordance with the edge normal vector. In the presented method, the descriptor is a vector of seven average pixel values in a given area. The size of the area includes the size of the edge detection mask that was selected for this case in the graph edge creation process. The size of the descriptor itself may vary depending on the solution. Longer descriptors map more details of the edge cross-section. However, it may also make it difficult to compare edges that are not perfectly similar to each other. Examples are parasitic eggs, whose cross sections can be stretched or narrowed.

4.1.2 Experimental Results

Experiments were performed on images extracted from video of microscopic examination. The video was captured from a camera attached to the microscope in the PAL resolution of 768×576 pixels.

Figure 4.9 shows application examples of the proposed method. Figure 4.9c presents the detected edges in the graph representation. We wanted the method to describe only edges constituting the outline, but in reality, it detects the inner edges as well. The biggest problem is highly transparent objects because their edges are less in contrast with the environment. In addition, in their case, there is often a phenomenon where their edges affect the refraction of light, creating a variety of reflections in the form of an example light halo effect. In their case, the method detects many near edges. Due to the principles of human perception, this problem is well illustrated by Fig. 4.9b, which represents the response of the edge detector mask. In this worse case, because of the transparency, the course of the edge is also often interrupted by elements located below or above the examined element. Unfortunately, comparing

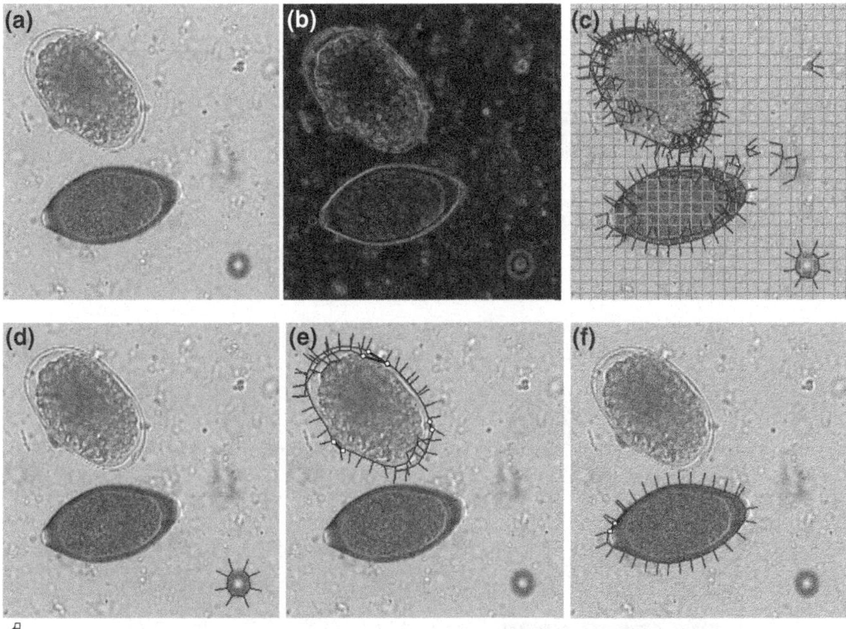

Fig. 4.9 Results obtained using edge descriptor and curvature coefficients. **c** Presents the entire graph with the adjacent grid. **d, e, f** Present resulting objects created by fragment grouping

the curvature of the edge alone in their case does not produce a good result, and only adding edge descriptors improves the results, eliminating internal edges. In addition to combining edges, the method can also group similar edges, which can be used to recognise them, especially since descriptors and coefficients can be normalised. Figure 4.10 shows two examples of searching for similar edges. In both cases, the method has drawn only matching edges, and the pattern edge that is used to search has been marked with a rectangle. In addition, descriptors of all line nodes were marked in both figures. As we can see from the descriptors marked in the figures, the descriptors for each node are not perfectly compatible. In the case of comparing each node separately, the result would be much worse since a large number of line nodes would not be grouped together. The proposed approach compares resultant descriptors of whole line fragments, so merged lines also take into account incompatible nodes previously connected by the edge segmentation and graphing process.

Descriptors depict edges only in grayscale. For other types of images, the proposed descriptors may describe the colour components separately. In the case of parasitic images, colour is less important and even often varies, depending on the type of sample illumination and the camera white balance setting.

Figure 4.11 presents more examples of the experimental results. The first column presents the results of graph creation presented in Table 4.1. The table presents numbers of graph line fragments and the total node numbers of all fragments. The

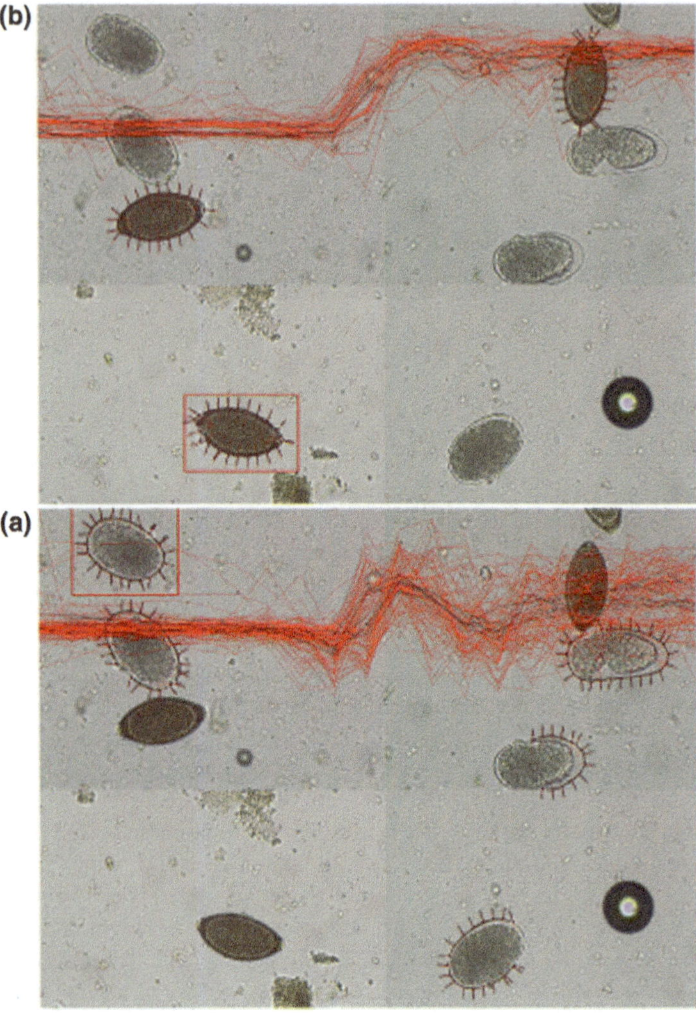

Fig. 4.10 An example of finding similar edges using only proposed descriptors. In the image there are drawn only similar fragments of graph, and the red rectangle marks the query image. Over the image, there is a layer added with a chart describing values of node descriptors (gradients). The red lines describe all node descriptors and black line describe average descriptor values for the objects

fourth column presents the time of graph creation during the edge detection, and the last column presents the total time of fragments search procedure for each line fragment. As we can see, the search for fragments is faster than the graph creation. Table 4.2 presents examples of fragments search results for Fig. 4.11. The second column presents the number of nodes of the search pattern used to performs the search. The third column presents the total number of nodes of the found fragments

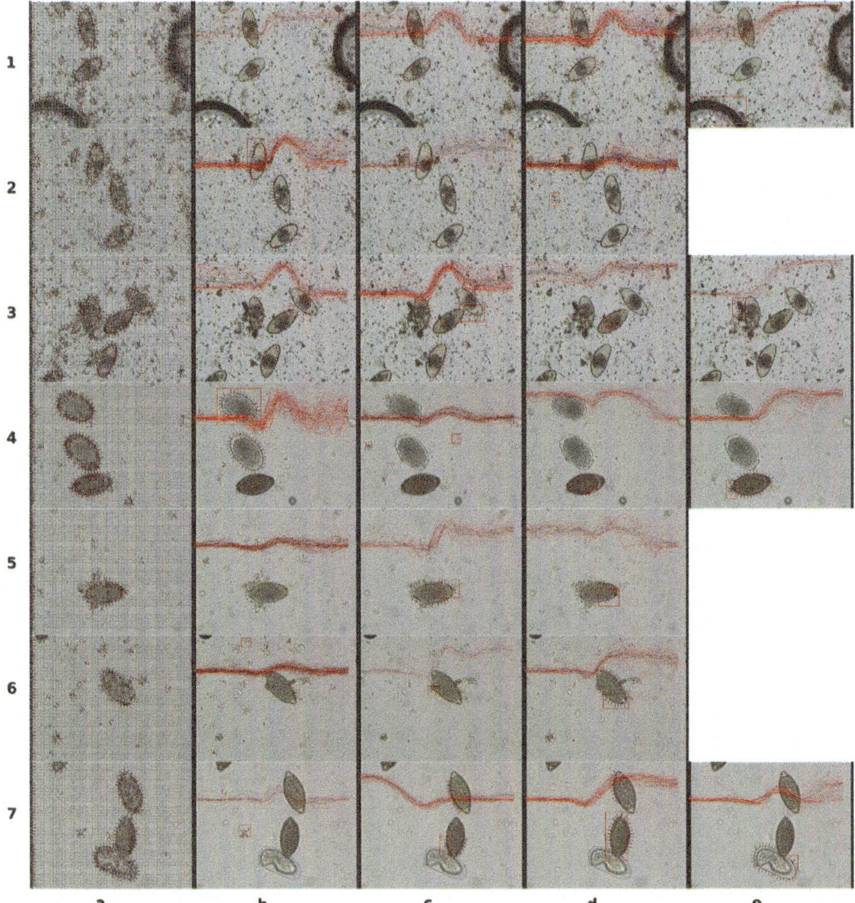

Fig. 4.11 Results of the experiment performed on the selected image. The first column presents the graph representation of the detected edges. The next columns present examples of finding similar edges by descriptors like in Fig. 4.10. Table 4.1 contains summary outcomes for these examples

which number is in the second column. The last column describes the time of the search for fragments without image drawing.

4.1.3 Conclusions

The advantage of the presented solution is that image edges are represented as graphs. The edge descriptors are generated only for nodes which is much more economical than the pixels constituting real edge points. In addition, because of the relation-

Table 4.1 Results of graphs creation for Fig. 4.11

Image	Nodes	Line fragments (s)	Graph make time	Groups check time (s)
1a	885	266	0.28	0.020
2a	747	223	0.28	0.017
3a	878	258	0.30	0.020
4a	647	146	0.11	0.007
5a	259	75	0.11	0.003
6a	298	82	0.11	0.003
7a	480	88	0.07	0.004

Table 4.2 Results of finding similar fragments by descriptors for Fig. 4.11

Image	Pattern nodes	All found nodes	Found fragments	Search time (ms)
1b	3	18	3	0.072
1c	5	47	4	0.071
1d	15	74	8	0.076
1e	21	42	7	0.064
2b	8	61	8	0.063
2c	4	20	6	0.060
2d	4	76	27	0.083
3b	3	42	5	0.068
3c	6	65	8	0.072
3d	3	27	6	0.066
3e	6	12	3	0.069
4b	31	73	4	0.042
4c	2	38	14	0.055
4d	2	28	3	0.044
4e	5	37	5	0.047
5b	2	44	15	0.042
5c	6	24	4	0.026
5d	6	24	4	0.026
6b	3	58	19	0.062
6c	2	7	2	0.021
6d	7	30	5	0.022
7b	3	11	3	0.028
7c	4	29	5	0.029
7d	17	42	3	0.028
7e	3	41	3	0.029

ships formed between the nodes, comparing graph path fragments comes down to a comparison of only two nearest ones instead of comparing all nodes.

Curvature coefficient determined for the entire part of the track allows pre-assess the degree of similarity between combinations of the edges. Moreover, in the case of including a new path or individual nodes, we do not need to determine the coefficients from scratch, because they can be added to the existing coefficients.

The disadvantage of the proposed solution could be revealed in the case of edges of small objects, which cannot be represented with enough accuracy by the proposed graph method. In this case, we can reduce the space between the nodes of the graph, but then the graph that represents edges will be suitably expanded, resulting in slower operation.

4.2 Interest Point Localization Based on the Gestalt Laws

This section proposes a method for grouping fragments of contours of objects in the images of microscopic parasitological examinations, characterised by the high transparency of analysed objects [11]. The method is based on a graphical representation of the edges in vector form, allowing to reduce the required calculations substantially. The method uses simple vector operations to determine stroke parameters describing the degree of curvature and closure of the examined contour and the direction where the remaining part of the contour should be located. Compared with other methods of detecting elliptic objects, the proposed method allows for the association of objects with rather irregular and distorted shapes occurring in parasitic images.

Detecting and defining object contours is particularly important from the point of view of object recognition methods as it allows to extract the shape of the object, which can then be compared with another shape [6, 16, 22]. These methods are quite commonly used being further development of edge detection methods. In the case of edge detection, only the boundary lines between the contrasting areas are searched. Most often, this is accomplished by applying masks that approximate the value of the pixel values in at least two orthogonal directions [7, 23]. The second most common method is image segmentation which defines areas of a similar pattern formed by neighbouring pixels [1, 8]. After segmentation, the edges of the image segments are automatically determined. When defining an object's contour, the problem of determining and selecting the edges that are genuine parts of the object's contour plays an important role. One of the problems is the edges separating the internal structure of objects. This is a significant problem, in particular, because in the case of real-life images, many objects are similar to the background, e.g. animals even try to mimic their surroundings. The methods defining the outline of the objects try to select the edges or, in the case of segmentation, try to connect the segments into larger groups. Another problem is the lack of detected edges in places where the object has a border, however, merges with the background. In the case of natural images, they are often shade areas that weaken the contrast between the edges or places where the subject and background have the same colour and texture.

Another aspect that is important for contour definition and image analysis methods is the evaluation of their quality. It is hard for men to analyse large numeric datasets. In the case of image analysis methods, the situation is very different. The man is equipped with a perfect mechanism of visual perception. As a result, even large data can be easily interpreted by humans if presented graphically. For this reason, it is complicated to develop a computational method that fully reflects the effects of human operator work. The problem stems from the fact that image analysis methods operate on numeric features and data. Human perception, on the other hand, represents an image in an interpreted form in which the man sometimes sees elements that do not actually exist. Human perception is described partially by the Gestalt laws of grouping [4, 15], which through graphic representations allow us to understand how differently people perceive the image in relation to reality. Some of these rules have been reflected in implementations of certain computer vision methods, most often in the detection, grouping, and refinement of the image [14]. From the point of view of the methods that have been developed so far and the possibilities of implementation, the most interesting Gestalt laws are:

- good continuation—indicating that in the case of encountered crossed edges, perception follows the one that least changes the current direction.
- closure—indicating that perception strongly determines contour to create a closed area, but also allows to specify multiple areas that have common edge paths.
- similarity—indicating that perception follows the edges even fragmented whose local gradient is most similar to current edge. However, the path can be also created by similar objects or even entire areas.
- continuity—human perception omits gaps in edge path. Even in the case of obscured objects, perception tries to look for further parts on the other side of the obscuring object. Perception also gives a feeling to the person that the object must have its continuation and creates an artificial image of the approximate shape.

These rules overlap each other and have been formed in the process of evolution, helping humans function in nature.

4.2.1 Problem Description

Coproscopic examinations are non-invasive diagnostic tests for parasitic diseases in animals and humans. They consist in microscopic analysis of samples taken from the patient to search for parasite eggs (Fig. 4.1) or live specimens [5, 18, 19, 24]. Compared to histopathological examinations [2, 5], the samples tested are an irregular suspension containing any amount of impurities of any shape (Fig. 4.1). The parasite eggs themselves are not ideal objects; they differ in their deformities and state of internal development. As mentioned earlier, in the case of determining the outline, the problem is the internal and adjacent structures of the image edges. In the case of faecal examination images, there are slightly different problems than in the case of general images and photographs. The most significant are:

- Transparency of objects. Transparency causes the problem of a rather blurry border between the object and the background. An additional problem is caused by contamination near the transparent edges. Their structure is generally much less transparent than the surrounding environment, giving a stronger response from edge detectors.
- Small differences between the colours and brightness of the different elements of the image. The cause is very thin samples that become transparent. The colour and contrast of the image are influenced by the illumination of the sample itself, causing the same samples to look different depending on the lighting and the camera used.
- Blur of the image fragments. The blur is caused by a large sample magnification. The high focal length results in a drastic decrease in depth of field. At first glance, the elements appear to be flat; however, their height exceeds the multiple of the depth of field. The problem, in this case, is the sharpness of the contours of the objects as they may be at different heights and in the part of the object to be blurred (Fig. 4.13). With the adjustment of the height of the sample the internal structure of the object also changes. In a sense, it can be said that the cross-section of this object is visible, and adjusting the height, we see further sections analogically to computer tomography. This phenomenon can also be used by a technician to evaluate better a sample being tested (Fig. 4.12).

4.2.2 Method Description

We developed a method that allows to associate edge fragments on the basis of similarity and to determine the location of the centres of the surrounding areas. This problem can be solved by several existing solutions, but in this case, we wanted to develop a method that simulates the way the human vision works. We aimed to

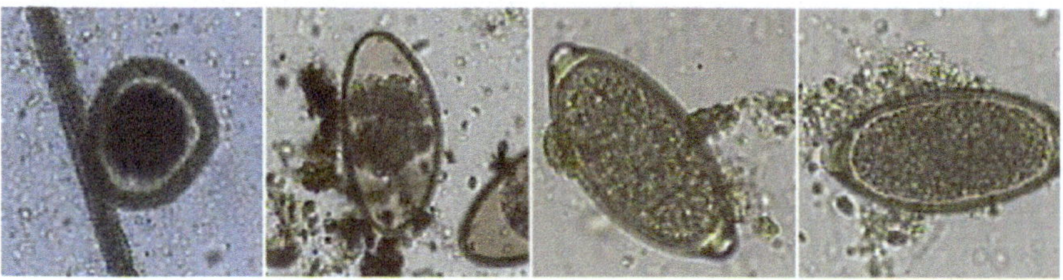

Fig. 4.12 Examples of parasite eggs of roundworm, pinworm and whipworm with impurities

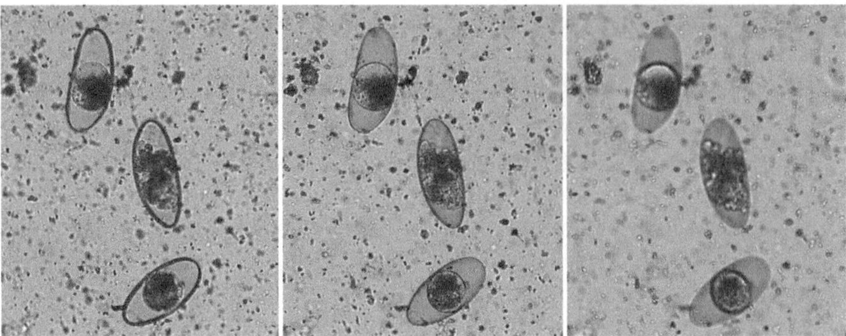

Fig. 4.13 Changing the structure of objects along with the horizontal position of the sample in the microscope

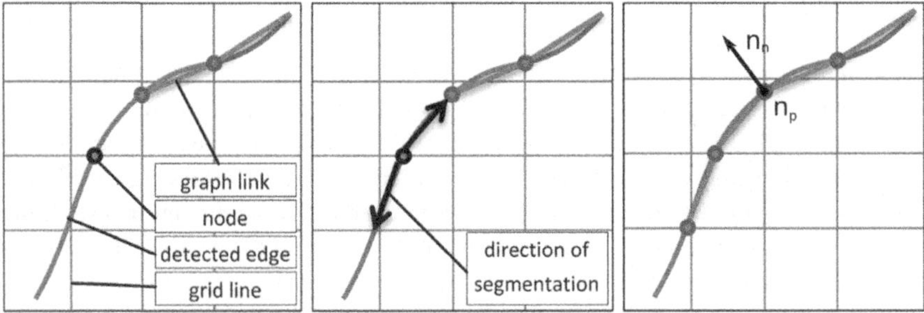

Fig. 4.14 Layout of graph nodes along the edge line

determine points of interest to which the man analysing the picture unintentionally focuses his or her attention, omitting the rest of the image. As mentioned earlier, the man has an excellent sense of image analysis what was partly characterised by the Gestalt laws; therefore they were decided to be reflected in the presented method.

4.2.2.1 Graph Representation of Image Edges

Firstly, we proposed to represent image edges as graphs [10], where image nodes are points along the detected image edge (Fig. 4.14).

For better results, the graph should have a uniform distribution of nodes. For this reason, the auxiliary grid was used in the method (Fig. 4.15). Graph nodes are defined at points of intersection of grid lines and edges of objects. In this way, the resulting graph has a uniform distribution, and the nodes are spaced by a distance measured in pixels so that the next steps are based on much less data that needs to be analysed as in the case of a bitmap with the pixel-marked edges. Detecting the edges runs initially along the grid lines so that most of the image will not have to be analysed, it is only important that the grid size is selected so that the searched objects are crossed

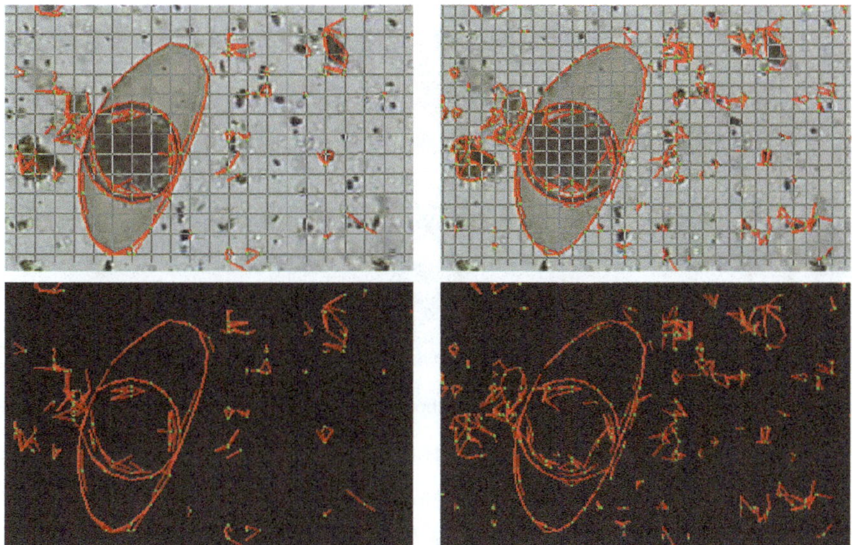

Fig. 4.15 Example of using the auxiliary grid. The difference between the number of the detected edges for two different grid sizes

at multiple points. An additional effect is to omit some of the contaminants that were not cut through the grid lines. Figure 4.15 illustrates two cases that use 15px and 20px grids, as we can see, fewer impurities are mapped for a lower density mesh.

The next step is to determine the edge of the graph, which is done by segmenting the edges of the image in opposite directions from the node found. Usually, two segments are created as a result of edge segmentation. The graphical representation is the generalization of data describing the edges of an image from a raster spatial representation to a vector point list. Each detected node is eventually described by a position vector $\mathbf{n}_p$ and normal vector $\mathbf{n}_n$ determining edge direction. Application of normal directional vectors $\mathbf{n}_n$ allows to easily examine the relationships between other vectors, limiting mainly to operations:

- summing vectors to obtain the resulting vector,
- dot product defining the cosine value of the angle between the vectors,
- cross product, on the basis of which a perpendicular vector can be determined.

Vector operations are commonly used in computer graphics, and most graphics accelerators are naturally suited to compute them.

4.2.2.2   Node Descriptor

Each node also has a local gradient descriptor so that it can be compared on the basis of their similarity. Figure 4.16 shows the visualisation of sample descriptors of

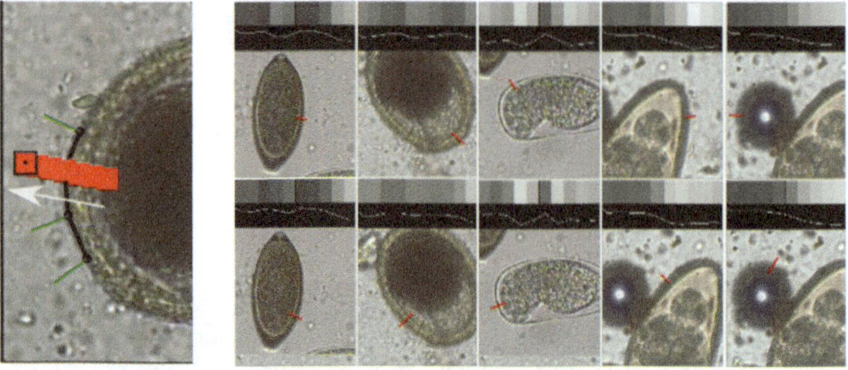

Fig. 4.16 Edge descriptors describing the neighbourhood of the graph nodes

edge fragments and a way to compute them. Each descriptor consists of a sequence of numbers that are average pixels along the edge. The cross-section direction is a normal vector $\mathbf{n}_n$ of the node determined during edge detection. Comparison of descriptors is done by examining the deviations between the values of both sets of values, where the descriptors are closer to each other when this value is smaller. The use of the descriptor allows the pairs of edges to reject ones that are definitely unlike each other when pairing edges. The descriptor reflects the principle of similarity and good continuation of the Gestalt laws.

4.2.2.3 Graph Path Descriptor

Once the graph is created, the method selects single paths. These paths include nodes that have only two edges of the graph, and boundary nodes are nodes having one or more edges of the graph. In addition, the method examines the nodes and divides the path if the node descriptors will be different, or if the difference in angle between the nodes' normal vectors will be significantly higher. During path isolation, the method determines a descriptor for each path, consisting of:

- The value c is the average value of the angle cosine between the normal vector of the node and the direction vector pointing to the next node.
- The resultant vector $\mathbf{n}$ being the vector of the sum of normal vectors $\mathbf{n}_n$.
- The resultant vector $|\mathbf{n}|$ which is a vector of sums of normal vector $\mathbf{n}_n$ modules.
- Vectors $\mathbf{n}_1$ and $\mathbf{n}_2$ describing the bounding box of the path.

The value of $\mathbf{n}$ describes the path bend and can take a value between -1 and 1. The value $c < 0$ indicates that the path is bent toward the normal vectors, $c = 0$ that is straight, $c > 0$ bent toward positive nodes. The resulting vector $\mathbf{n}$ determines the direction in which the contour is missing. Length of the vector $\|\mathbf{n}\|$ in the case of fully closed objects should be close to 0, since node vectors on both sides will neutralize

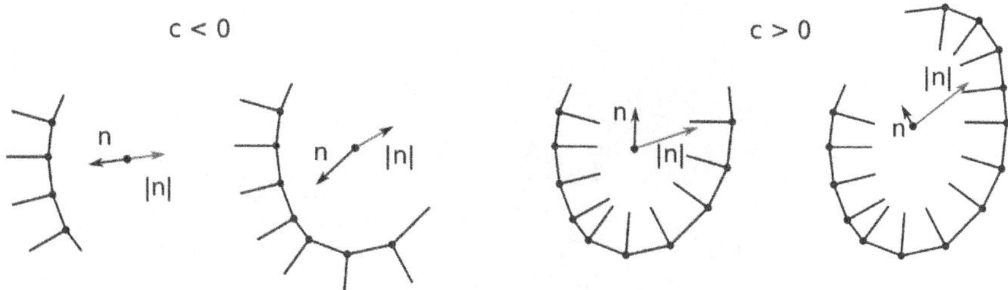

Fig. 4.17 Examples of graph path with associated vectors **n** and |**n**|

the resultant vector **n**, otherwise it will indicate the direction in which there is lack of closure. In the case $c < 0$ this vector will indicate the opposite direction.

Vector **n** indicates the number of nodes and together with the bounding vectors $\mathbf{n}_1$ and $\mathbf{n}_2$ allows to determine if the number of nodes is sufficient to create a closed contour in the designated area.

4.2.2.4 Determining the Points of Interest

At this stage we have the edge path determined and described thus we can proceed to determine interest points. To this end, the method compares two fragments. First, we compare bend values of c fragments, and then node descriptors. If they are not the same, the connection is rejected. Then, the method checks to see if the combination of fragments improves the closure of the outline by summing the vectors **n** and |**n**|. If the length of the vector **n** increases, it means that the closure is degraded and the pair is rejected. It is then examined whether the paths are curved towards each other on the basis of the determined angle cosine between vector n and the vector defining the direction of the second path. Finally, node descriptors are also compared. If the paths are consistent there is a point of interest consisting of two paths and new values **n**, |**n**| and the bounding box ($\mathbf{n}_1$, $\mathbf{n}_2$). In the next steps, the new point of interest is checked with the already existing ones. If it is consistent with the bounding box and their combination reduces the length of the vector n, then the new point is appended to the existing one. The last step is the point of interest evaluation. The method analyzes each point by checking the closure factor. In the first place, the method checks the vector length coefficient **n** to |**n**| which ranges between 1 and 0, where values close to 0 indicate a better closure of the outline. Then the method divides bounding box circumference by the grid size, thus obtaining the approximate number of nodes that should have a point of interest to create a closed area. Then, this number is compared with the vector length |**n**|. In this way the method is able to reject points of interest that are too distinctive and could not create coherent areas because of the lack of nodes.

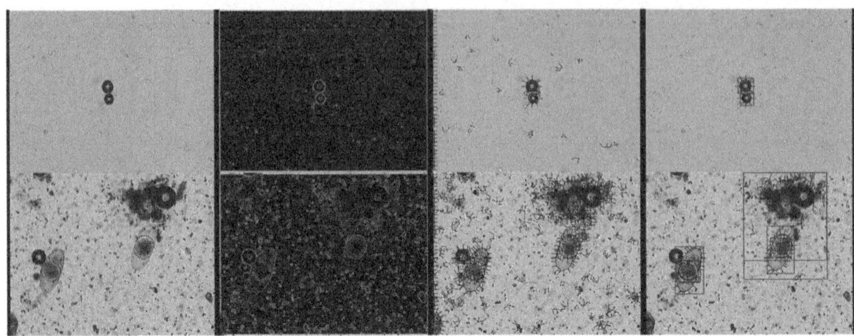

Fig. 4.18 Experimental results. The first column contains original images. The second column contains edge detection results, the third one—the entire graph representation of the image edges and finally, the area of interest determined from the graph structure

Table 4.3 The results of experiments for the graph created from Fig. 4.17

Image	Number of					
	Edge pixels	Graph nodes	Graph paths	Compared paths	Matched paths	Points of
1	42549	142	29	41	11	2
2	121917	874	246	6730	1606	7

4.2.3   Experiments

Experimental studies were conducted on images extracted from video records documenting process of microscopic samples diagnosis. The extracted images are video frames with a resolution of 768×576 pixels. The content of the images presents a random mix of non-cropped objects, usually parasites eggs, food contamination or air bubbles. An example of experimental results is presented in Fig. 4.18. Parameters of the generated graphs are presented in Table 4.3 and the detected areas of interest are described in Table 4.4. The graph was created with the grid size 15px, this means that each node maps and represents up to approximately 15px of the picture edge. In addition, the grid usage allowed to omit some of the undesirable objects, which is best presented in the second example image.

In addition, the applied grid allowed to omit some of the unwanted pollutions. The second column of Table 4.3 describes the total number of edge pixels that have been mapped by graph node (the third column) which number is approximately 0.7% of the number of pixels. The fourth column shows the number of standalone paths from the graph that are compared to find the closed areas. Table 4.4 presents the determined points of interest, which number depends on the image content. In some cases, the points of interest are duplicated because of a different combination of compared paths. Outline of the interest area described by graph paths allows representing their readable shape in the vector form using a relatively small number of points compared to their size.

Table 4.4 The list of detected interest points for Fig. 4.18

Image	Position	Radius	Number of	
			Graph nodes	Graph paths
1	370,270	21	11	1
	370,270	18	9	1
2	511,155	146	85	15
	512,189	163	90	18
	497,264	30	14	2
	160,346	32	15	2
	496,272	60	29	1
	160,348	35	13	2
	164,344	61	30	2

4.2.4 Conclusions

The presented method allows detecting points of interest by a gradual generalisation of data from the edge pixels by graph nodes to paths. Thanks to that, the amount of the required comparisons has been drastically reduced. Using a grid-based graph creation method in the case of microscopic images allowed to omit most of the small unwanted objects at the beginning of the whole image processing. However, there are images where removing these details may be a problem. The proposed method allows to detect and describe areas that have an irregular outline with missing fragments, but this leads to a problem, as the method can detect erroneous redundant areas. The points of interest or graphs found by the presented method can be used by other methods such as the active shape model. The application can be various types of images.

4.3 Summary and Discussion

The chapter presented methods for describing edges. The first method reconstructs connections between parts of object outlines in images. Examined objects in microscopic images are highly transparent; moreover, close objects can overlap each other. Thus, the segmentation and separation of such objects can be difficult. Another frequently occurring problem is partial blur due to high image magnification. Large focal length of a microscope dramatically narrows the depth of field). The most important advantage of the presented solution is that image edges are represented as graphs. The edge descriptors are generated only for nodes which is much more economical than the pixels constituting real edge points. Besides, because of the relationships formed between the nodes, comparing graph path fragments comes down to a comparison of only two nearest ones instead of comparing all nodes.

The next method proposed a method for grouping fragments of contours of objects in the images of microscopic parasitological examinations, characterised by the high transparency of analysed objects. The method is based on a graphical representation of the edges in vector form, allowing to reduce the required calculations substantially. The method uses simple vector operations to determine stroke parameters describing the degree of curvature and closure of the examined contour and the direction where the remaining part of the contour should be located. Compared with other methods of detecting elliptic objects, the proposed method allows for the association of objects with rather irregular and distorted shapes occurring in parasitic images. The points of interest or graphs found by the presented methods can be used by various computer vision systems and can be applied to be various types of images.

References

1. Achanta, R., Shaji, A., Smith, K., Lucchi, A., Fua, P., Süsstrunk, S., et al.: Slic superpixels compared to state-of-the-art superpixel methods. IEEE Trans. Pattern Anal. Mach. Intell. **34**(11), 2274–2282 (2012)
2. Cireşan, D.C., Giusti, A., Gambardella, L.M., Schmidhuber, J.: Mitosis detection in breast cancer histology images with deep neural networks. In: International Conference on Medical Image Computing and Computer-assisted Intervention, pp. 411–418. Springer (2013)
3. Flores-Quispe, R., Velazco-Paredes, Y., Escarcina, R.E.P., Castañón, C.A.B.: Automatic identification of human parasite eggs based on multitexton histogram retrieving the relationships between textons. In: 2014 33rd International Conference of the Chilean Computer Science Society (SCCC), pp. 102–106. IEEE (2014)
4. Guberman, S., Maximov, V.V., Pashintsev, A.: Gestalt and image understanding. Gestalt Theor. **34**(2), 143 (2012)
5. Jiang, M., Zhang, S., Huang, J., Yang, L., Metaxas, D.N.: Scalable histopathological image analysis via supervised hashing with multiple features. Med. Image Anal. **34**, 3–12 (2016)
6. Lim, J.J., Zitnick, C.L., Dollár, P.: Sketch tokens: A learned mid-level representation for contour and object detection. In: Proceedings of the IEEE Conference on Computer Vision and Pattern Recognition, pp. 3158–3165 (2013)
7. Marr, D., Hildreth, E.: Theory of edge detection. Proc. R. Soc. Lond. B **207**(1167), 187–217 (1980)
8. Martin, D.R., Fowlkes, C.C., Malik, J.: Learning to detect natural image boundaries using local brightness, color, and texture cues. IEEE Trans. Pattern Anal. Mach. Intell. **26**(5), 530–549 (2004)
9. Milborrow, S., Nicolls, F.: Locating facial features with an extended active shape model. Comput. Vis. ECCV **2008**, 504–513 (2008)
10. Najgebauer, P., Nowak, T., Romanowski, J., Rygal, J., Korytkowski, M.: Representation of edge detection results based on graph theory. In: Rutkowski, L., Korytkowski, M., Scherer, R., Tadeusiewicz, R., Zadeh, L.A., Zurada, J.M. (eds.) Artificial Intelligence and Soft Computing, pp. 588–601. Springer, Berlin, Heidelberg (2013)
11. Najgebauer, P., Rutkowski, L., Scherer, R.: Interest point localization based on edge detection according to gestalt laws. In: 2017 2nd IEEE International Conference on Computational Intelligence and Applications (ICCIA), pp. 349–353 (2017)
12. Najgebauer, P., Rutkowski, L., Scherer, R.: Novel method for joining missing line fragments for medical image analysis. In: 2017 22nd International Conference on Methods and Models in Automation and Robotics (MMAR), pp. 861–866 (2017)

13. Nowak, T., Najgebauer, P., Ryga, J., Scherer, R.: A novel graph-based descriptor for object matching. In: Rutkowski, L., Korytkowski, M., Scherer, R., Tadeusiewicz, R., Zadeh, L., Zurada, J. (eds.) Artificial Intelligence and Soft Computing. Lecture Notes in Computer Science, vol. 7894, pp. 602–612. Springer, Berlin, Heidelberg (2013)
14. Papari, G., Petkov, N.: Adaptive pseudo dilation for gestalt edge grouping and contour detection. IEEE Trans. Image Process. **17**(10), 1950–1962 (2008)
15. Petermann, B.: The Gestalt Theory and the Problem of Configuration. Routledge (2013)
16. Ren, X., Fowlkes, C.C., Malik, J.: Scale-invariant contour completion using conditional random fields. In: Null, pp. 1214–1221. IEEE (2005)
17. Rogers, M., Graham, J.: Robust active shape model search. In: European Conference on Computer Vision, pp. 517–530. Springer (2002)
18. Suzuki, C.T., Gomes, J.F., Falcao, A.X., Papa, J.P., Hoshino-Shimizu, S.: Automatic segmentation and classification of human intestinal parasites from microscopy images. IEEE Trans. Biomed. Eng. **60**(3), 803–812 (2013)
19. Tchiotsop, D., Tchinda, R., Didier, W., NOUBOM, M.: Automatic recognition of human parasite cysts on microscopic stools images using principal component analysis and probabilistic neural network. Editor. Pref. **4**(9) (2015)
20. Tek, F.B., Dempster, A.G., Kale, I.: Computer vision for microscopy diagnosis of malaria. Malar. J. **8**(1), 153 (2009)
21. Veta, M., Van Diest, P.J., Willems, S.M., Wang, H., Madabhushi, A., Cruz-Roa, A., Gonzalez, F., Larsen, A.B., Vestergaard, J.S., Dahl, A.B., et al.: Assessment of algorithms for mitosis detection in breast cancer histopathology images. Med. Image Anal. **20**(1), 237–248 (2015)
22. Wang, S., Kubota, T., Siskind, J.M., Wang, J.: Salient closed boundary extraction with ratio contour. IEEE Trans. Pattern Anal. Mach. Intell. **27**(4), 546–561 (2005)
23. Yang, K., Gao, S., Li, C., Li, Y.: Efficient color boundary detection with color-opponent mechanisms. In: Proceedings of the IEEE Conference on Computer Vision and Pattern Recognition, pp. 2810–2817 (2013)
24. Yang, Y.S., Park, D.K., Kim, H.C., Choi, M.H., Chai, J.Y.: Automatic identification of human helminth eggs on microscopic fecal specimens using digital image processing and an artificial neural network. IEEE Trans. Biomed. Eng. **48**(6), 718–730 (2001)

Chapter 5
Image Retrieval and Classification in Relational Databases

Relational databases are used to store information in every kind of life and business. They are suited for storing structured data and binary large objects (BLOBs). Unfortunately, BLOBs and multimedia data are difficult to handle, index, query and retrieve. Usually, relational database management systems are not equipped with tools to retrieve multimedia by their content. One of the seminal solutions devoted to the storage and retrieval of images in a database was presented in [18]. They compared images in PostgreSQL by colour. In [28] User-Defined Functions (UDFs) were used in order to find duplicate images. There were also attempts to implement search methods for images in large data sets in commercial database systems. An example might be the Oracle database environment called "interMedia", where image retrieval is based on the global colour representation, low-level patterns and textures within the image, such as graininess or smoothness, the shapes that appear in the image created by a region of uniform colour and their location.

New methods for searching for images directly from a database require to extend the standard SQL language for content-based image retrieval commands. The authors of [1, 19] proposed a query language similar to a full text search engine. Multimedia XML-like language for retrieving multimedia on the Internet was proposed in [17]. As aforementioned, standard SQL does not contain commands for handling multimedia, large text objects and spatial data. Thus, communities that create software for processing such specific data types began to draw up SQL extensions, but they transpired to be incompatible with each other. That problem caused abandoning new task-specific extensions of SQL and a new concept won, based on libraries of object types SQL99 intended for processing specific data applications. The new standard, known as SQL/MM (full name: SQL Multimedia and Application Packages), was based on objects, thus programming library functionality is naturally available in SQL queries by calling library methods. SQL/MM consists of several parts: framework—library for general purposes, full text—defines data types for storing and searching a large

© Springer Nature Switzerland AG 2020
R. Scherer, *Computer Vision Methods for Fast Image Classification
and Retrieval*, Studies in Computational Intelligence 821,
https://doi.org/10.1007/978-3-030-12195-2_5

amount of text, spatial—for processing geospatial data, still image—defines types for processing images and data mining—data exploration. There are also attempts to create some SQL extensions using fuzzy logic for building flexible queries. In [4, 8, 9] possibilities of creating flexible queries and queries based on users examples are presented. It should be emphasized that the literature shows little efforts of creating a general way of querying multimedia data.

Popular way of dealing with a high number of local features generated to describe images is the bag-of-features (BoF) approach [7, 14, 20, 24, 30, 31] has gained in popularity. In the BoF method, clustered vectors of image features are collected and sorted by the count of occurrence (histograms). All individual descriptors or approximations of sets of descriptors presented in the histogram form must be compared. The information contained in descriptors allows for finding a similar image to the query image. Such calculations are computationally expensive. Moreover, the BoF approach requires to redesign the classifiers when new visual classes are added to the system.

All these aforementioned methods require a large amount of data and computing power to provide an appropriate efficiency. Despite applying some optimization methods to these algorithms, the data loading process is time-consuming. In the case of storing the data in the database, when a table contains n records, the similarity search requires $O(log_2 n)$ comparisons. Image comparison procedure can take less time when some sorting mechanisms are applied in a database management system. Noteworthy solutions are proposed by different database products [3, 13, 25].

This chapter describes a family of solutions for indexing and retrieving images in relational database management systems (RDBMSs). The methods use local image features to describe images.

5.1 Bag of Features Image Classification in Relational Databases

In this section, we present a novel relational database architecture aimed to visual objects classification [12]. The framework is based on the bag-of-feature image representation model [7, 20, 24, 31] combined with the Support Vector Machine classifier and is integrated with a Microsoft SQL Server database. Storing huge amount of undefined and unstructured binary data and its fast and efficient searching and retrieval is the main challenge for database designers. Examples of such data are images, video files etc. Users of the world most popular relational database management systems (RDBMS) such as Oracle, MS SQL Server and IBM DB2 Server are not encouraged to store such data directly in the database files. The example of such an approach can be Microsoft SQL Server where binary data is stored outside the RDBMS, and only the information about the data location is stored in the database tables. MS SQL Server utilises a special field type called FileStream which integrates SQL Server database engine with NTFS file system by storing large binary object

(BLOB) data as files in the file system. Transact-SQL (Microsoft SQL dialect) statements can insert, update, query, search, and back up FileStream data. The application programming interface provides streaming access to the data. FileStream uses operating system cache for caching file data. This helps to reduce any negative effects that FileStream data might have on the RDBMS performance. Filestream data type is stored as a `varbinary(max)` column with the pointer to actual data which is stored as BLOBs in the NTFS file system. By setting the FileStream attribute on a column and consequently storing BLOB data in the file system, we achieve the following advantages:

- Performance is the same as the NTFS file system and SQL Server cache is not burden with the Filestream data,
- The standard SQL statements such as SELECT, INSERT, UPDATE, and DELETE work with FileStream data; however, associated files can be treated as standard NTFS files.

In the proposed system, large image files are stored in a FileStream field. Unfortunately, despite using this technique, there does not exist a technology for fast and efficient retrieval of images based on their content in existing relational database management systems. Standard SQL language does not contain commands for handling multimedia, large text objects, and spatial data.

We designed a special type of field, in which a set of keypoints can be stored in an optimal way, as so-called User-Defined Type (UDT). Along with defining the new type of field, it is necessary to implement methods to compare its content. When designing UDT, various features must also be implemented, depending on implementing the UDT as a class or a structure, as well as on the format and serialisation options. This could be done using one of the supported .NET Framework programming languages, and the UDT can be implemented as a dynamic-link library (DLL), loaded in MS SQL Server. Another major challenge is to create a unique database indexing algorithm, which would significantly speed up answering SQL queries for data based on the newly defined field.

5.1.1 System Architecture and Relational Database Structure

Our system and generally the bag-of-words model can work with various image features. Here we use the SIFT features as an example. To calculate SIFT keypoints we used the OpenCV library. We did not use functions from this library as user-defined functions (UDF) directly in the database environment as they can be written only in the same .NET framework version as the MS SQL Server (e.g. our version of MS SQL Server was created based on .NET 4.0). Moreover, the calculations used to detect image keypoints are very complex, thus running them directly on the database server causes the database engine to become unresponsive (Fig. 5.1).

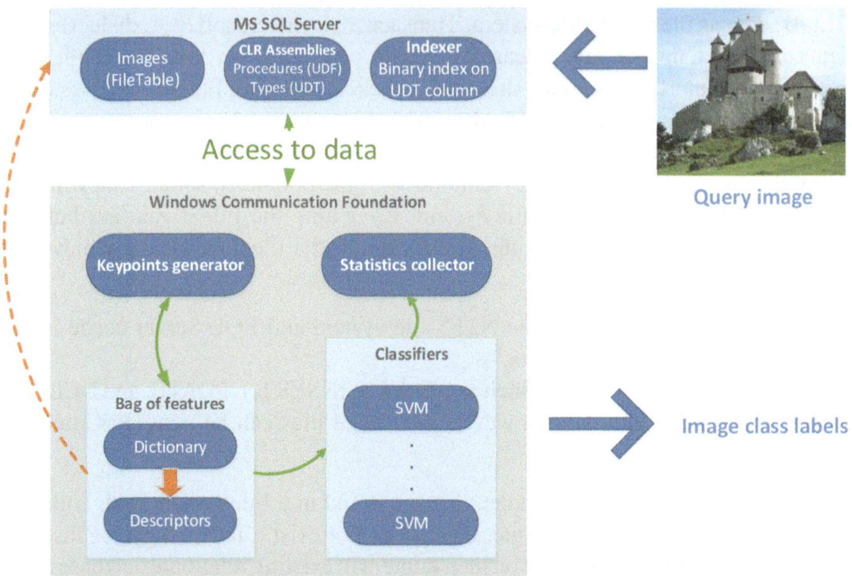

Fig. 5.1 Architecture of the proposed database image classification system

From the above-mentioned reasons, similarly as in the case of the Full-Text Search technology, the most time-consuming computations are moved to the operating system as background system services of WCF (Windows Communication Foundation). WCF Data Service works as the REST architecture (Representational State Transfer) which was introduced by Roy T. Fielding in his PhD thesis [6]. Thanks to the WCF technology, it is relatively easy to set the proposed solution on the Internet. To store image local keypoints in the database, we created a User-Defined Type (column `sift_keypoints` in `SIFTS` table). These values are not used in the classification of new query images. They are stored in case we need to identify a new class of objects in the existing images as having keypoint values, we would not have to generate keypoint descriptors again. The newly created type was created in C# language as a CLR class and only its serialised form is stored in the database (Fig. 5.2).

The database stores also Support Vector Machine classifiers parameters in the `SVMConfigs` table. Such an approach allows running any time the service with learned parameters. Running the service in the operating system will cause reading SVM classifiers from the database. The `Stats` table is for collecting algorithm statistics, where the most important numbers are execution times for consecutive stages of the algorithm. The `Images` table is for storing membership of images for visual classes. `Dictionaries` table is responsible for storing keypoint clusters data, and these cluster parameters are stored in the `DictionaryData` field of UDT type:

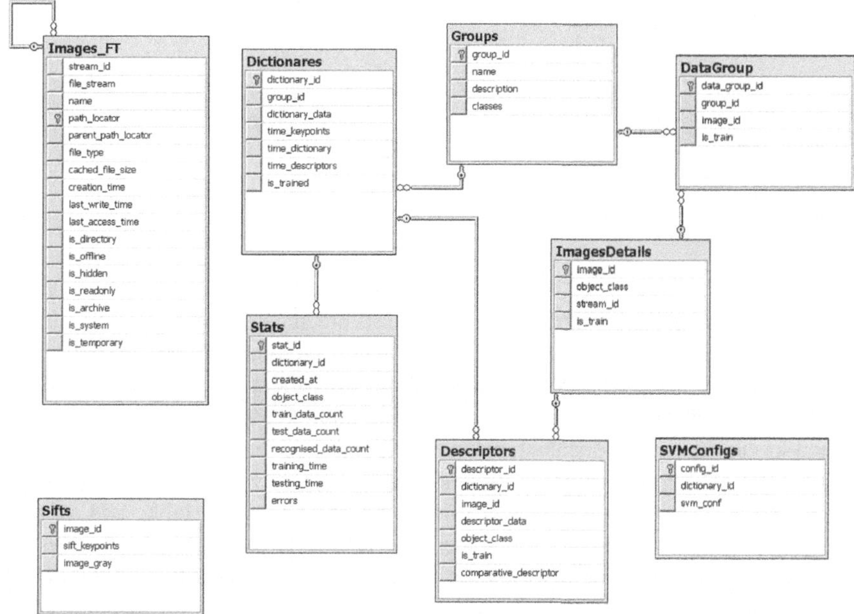

Fig. 5.2 A database structure of the proposed bag-of-feature retrieval system

```
public struct DictionaryData:
        INullable, IBinarySerialize
{
    private bool _null;
    public int WordsCount {get; set;}
    public int SingleWordSize {get; set;}
    public double[][] Values {get; set;}
    public override string ToString()
    ...
}
```

The `WordsCount` variable stores information about the number of words in the BoF dictionary, `SingleWordSize` variable value depends on the algorithm used to generate image keypoint descriptors, and in case of the SIFT algorithm, it equals 128. Two-dimensional matrix `Values` stores information regarding cluster centers. The system operates in two modes: learning and classification. In the learning mode, image keypoint descriptors are clustered to build a bag-of-features dictionary by the k-means algorithm. Cluster parameters are stored in `DictionaryData` variables. Next, image descriptors are created for subsequent images. They can be regarded as histograms of membership of image local keypoints to words from dictionaries. We use `SIFTDetector` method from the Emgu CV (http://www.emgu.com) library. Obtained descriptors are then stored in the `Descriptors` table of UDT type:

```
public struct DescriptorData:
            INullable, IBinarySerialize
{
    // Private member
    private bool _null;
    public int WordsCount {get; set;}
    public double[] Values {get; set;}
    ...
}
```

Using records from this table, learning datasets are generated for SVM classifiers to recognize various visual classes. These classifiers parameters are stored after the training phase in the `SVMConfigs` table. In the classification phase, the proposed system works fully automatically. After sending an image file to the `Images_FT` table, a service generating local interest points is launched. In the proposed approach, we use SIFT descriptors. Next, the visual descriptors are checked against membership to clusters stored in the database in the `Dictionaries` table, and on this base, the histogram descriptor is created. To determine membership to a visual class, we have to use this vector as the input for all SVM classifiers obtained in the learning phase. For the classification purposes, we extended SQL language and defined `GetClassOfImage()` method in C# language and added it to the set of User-Defined Functions. The argument of this method is the file identifier from the FileTable table.

Microsoft SQL Server constraints the sum of indexed columns to 900 bytes. Therefore, it was not possible to create an index on the columns constituting visual descriptors. To allow fast image searching of the `Descriptors` table, we created a field `comparative_descriptor` that stores descriptor value hashed by the MD5 algorithm [21]. It allowed creating the index on this new column; thus the time to find an image corresponding with the query image was reduced substantially.

5.1.2 Numerical Simulations

We tested the proposed method on three classes of visual objects taken from the PASCAL Visual Object Classes (VOC) dataset [5], namely: Bus, Cat and Train. We divided these three classes of objects into learning and testing examples. The testing set consists of 15% images from the whole dataset. Before the learning procedure, we generated local keypoint vectors for all images from the Pascal VOC dataset using the SIFT algorithm. All simulations were performed on a Hyper-V virtual machine with MS Windows Operating System (8 GB RAM, Intel Xeon X5650, 2.67 GHz). The testing set only contained images that had never been presented to the system during the learning process (Fig. 5.3).

The bag-of-features image representation model combined with the Support Vector Machine (SVM) classification was run five times for various dictionary sizes: 40,

Fig. 5.3 Example images from the test subset of the PASCAL Visual Object Classes (VOC) dataset [5]

Table 5.1 Classification accuracy for various sizes of the bag of words dictionary

Number of words in dict.:	40 (%)	50 (%)	80 (%)	100 (%)	130 (%)	150 (%)
Bus	40	50	60	60	70	50
Cat	90	80	50	80	80	80
Train	0	0	10	20	10	10
Total	43	43	40	53	53	47

50, 80, 100, 130 and 150 words. Dictionaries for the BoF were created using C++ language, based on the OpenCV Library [2]. The results of the BoF and SVM classification on the testing data are presented in Table 5.1. The SQL queries responses are nearly real-time for even relatively large image datasets.

5.1.3 Conclusions

We presented a method that allows integrating relatively fast content-based image classification algorithm with relational database management system. Namely, we used bag of features, Support Vector Machine classifiers and special Microsoft SQL Server features, such as User-Defined Types and CLR methods, to classify and retrieve visual data. Moreover, we created indexes to search for the same query image in large sets of visual records. Described framework allows automatic searching and retrieving images on the base of their content using the SQL language. The SQL responses are nearly real-time on even relatively large image datasets. The system can be extended to use different visual features or to have a more flexible SQL querying command set.

5.2   Bag of Features Image Retrieval in Relational Databases

In the previous section, we presented a system for image classification built in a relational databases management system. In this section, a database image retrieval system is shown [26], i.e. it returns a set of images similar to the query image which was presented on the system input. MS SQL Server offers the FileTable mechanisms thanks to the SQL Server Filestream technology to store large files a file system. Modifying the content of objects stored in a FileTable can be performed by adding, or removing data from directories linked to this table and the changes are visible in the table automatically. In the next subsections, the architecture and functionality of the proposed image retrieval system are presented.

5.2.1   Description of the Proposed System

The system described in this section allows searching similar images to the query image which was provided by a user or a client program. Users are able to interact with our system by executing a stored procedure. There is also a possibility of calling the methods from a WCF service in a remote way. This operation can be performed in client software. When the user interacts with the system locally, the query images can be copied to a special directory called `Test`, which is an integral part of the database FileTable structure. As a consequence, the appropriate trigger is executed and an adequate testing stored procedure is called. When client software connects to the system remotely, it is necessary to transfer the query image as a stream over the network. The authors provided API mechanisms to perform this kind of interaction.

5.2.1.1   Architecture of the System

The primary target of the system is business applications that need fast image retrieval functionality. It encapsulates computer vision algorithms and other mechanisms; thus the user does not have to know how to implement them. MS SQL Server 2012 provides the UDT mechanism (User-Defined Types) which was used for crucial elements such as image keypoints, dictionaries, or descriptors. All UDT types were programmed with custom serialization mechanisms. These types are stored in assemblies included in the database which is linked to our system. The software was based on .NET platform. Moreover, the additional advantage is the use of the Filestream technology which is included in MS SQL Server. As a consequence, reading high-resolution images is much faster than with using classical methods. The aforementioned technology provides the interaction with the image database, based on the content of appropriate folders (linked to FileTable objects), designed for storing images. Placing new images in these folders fires the adequate trigger. It gives the advantage of automatic initialization of the corresponding database objects without additional operations. Users have to indicate a query image to compare. As a result, the system returns the sequence of images similar to the content of the query image. The process of extending the set of indexed images in the database boils down to copying images to FileTable directories. Then, the dictionary and image descriptors are be generated automatically after inserting the number of words for the dictionary in an appropriate stored procedure. Figure 5.4 presents the architecture which was divided into four layers. In the first layer, the user selects a query image for transferring to the system over the remote WCF channel or by copying to the `Test` folder locally. After processing the query image, the user obtains the response as the sequence of similar images (sorted in descending order from the most similar image). The second layer is an interface which allows performing queries to the system database. The list of similar images consists of file paths from a database and similarity levels assigned to appropriate files. The third layer acts as the physical MS SQL Server database where the information about the images and their descriptors is stored. The table with descriptors is indexed to speed up generating the response. At this level, it is

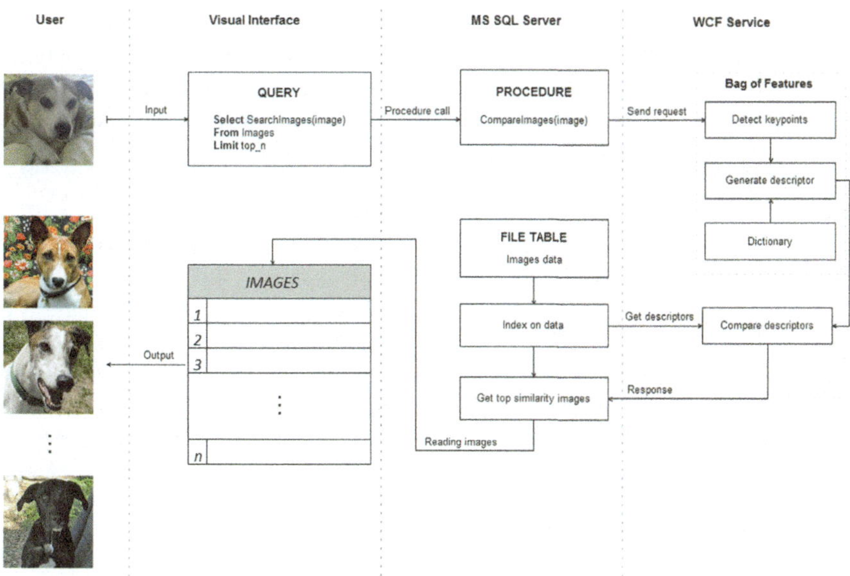

Fig. 5.4 Architecture of the proposed image retrieval system

also possible to execute a stored procedure which contributes to running the bag-of-features algorithm and indicating similar images over the WCF endpoint. The last layer contains the WCF service functionality. Methods shared by the web service module run the main algorithms, generate keypoints and descriptors basing on the dictionary. Having the dictionary, it is possible to perform the similarity calculation procedure. The response collected from the system contains a sorted list which is transferred to the second layer. The list stores top n most similar images, which can be accessed from the first layer.

5.2.1.2 System Functionality

The system is divided into modules, which are dedicated to specific functions. These modules include communication interfaces with other modules. The layered software implementation allows modifying some modules, without interfering with the other architecture parts of the system.

The domain model layer is a fundamental module for the business logic of the system and was created with the Database First approach. Figure 5.5 presents the database diagram. Considering the integration of the applied mechanisms from .NET platform, Microsoft SQL Server 2012 was chosen. The database structure was designed based on the bag-of-features algorithm. Keypoints, dictionaries and descriptors were stored in the database as UDT (User-Defined Types), for which serialization mechanisms were implemented. System functionality is mainly based on the bag-of-features algorithm because of its relatively high effectiveness and fast operation. Image features are calculated using the SIFT algorithm; nevertheless, the system can use other visual feature generation techniques. The local features calculated for images are stored in the database along with the dictionary structures and descriptors generated basing on these dictionaries. This approach entails the requirement of only one generation of crucial data structures for the system. The

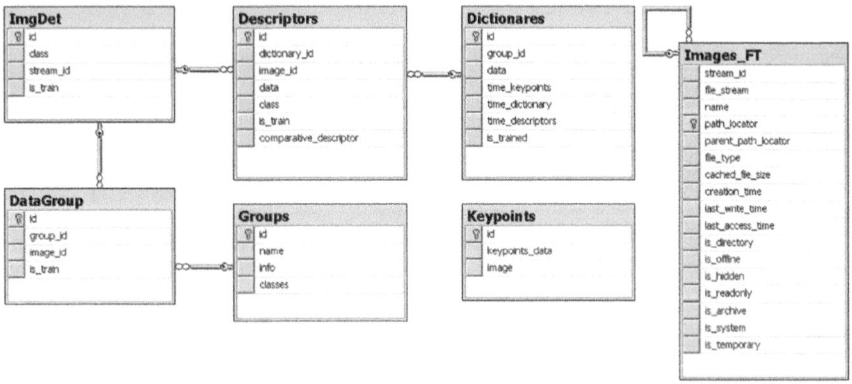

Fig. 5.5 Database diagram of the image retrieval system

`Images_FT` table was designed with the FileTable technology and contains images which are necessary for the training process. As a consequence, the entire content of this table influences clusters calculation and the effectiveness of image similarity detection.

Query by image operation relies on the initial dictionary loading with the appropriate identification number from the `Dictionaries` table. This operation is crucial for calculating descriptors for the adequate dictionary. The next procedure compares the query image descriptor with the other descriptors stored in the database. Vectors $\mathbf{x} = \{x_1, x_2, \ldots, x_n\}$ are generated for images from the database, and $\mathbf{y} = \{y_1, y_2, \ldots, y_n\}$ is calculated for the query image. The next procedure is responsible for comparing descriptors by the Euclidean distance. As a result, we determine the similarity factors for all comparisons sorted in descending order.

In an attempt to provide remote interaction with the system, we implemented an SOA layer (Service Oriented Architectures) in .NET technology. To achieve that essential aim, WCF (Windows Communication Foundation) web service was programmed. In this case, client software can execute procedures remotely. The system architecture also provides the distributed processing system, when a database server is situated in a different physical location. Hence, we implemented remotely executed WCF methods from stored procedures.

5.2.2 Numerical Experiments

In this section, we present the results of example experiments performed to validate the correctness of the system. We used images taken from the PASCAL Visual Object Classes (VOC) dataset [5]. We queried the database with images, and the returned images are shown with the distances to the query descriptor.

The first part of the tests was performed for query images which were not included in the database. When an image is presented on the system input, the response vector $R = S(x, y1), S(x, y2), \ldots, S(x, yN)$ obviously did not include similarity values being equal zero. It contained k similar images from an appropriate class of the query image. Figure 5.6 presents an example with several returned images. The next experiments consisted in showing images which had the exact representation in the database (in `Images_FT` table), i.e. they were included in the dictionary building process. In this case, the response vector included the output with m values equal zero when m indicates the number of identical images contained in the database. If the request was configured for including the k similar images, when $k > m$, then response vector should comprise $k > m$ values greater than zero. Figure 5.7 shows an example of querying the database with an image that existed in the database.

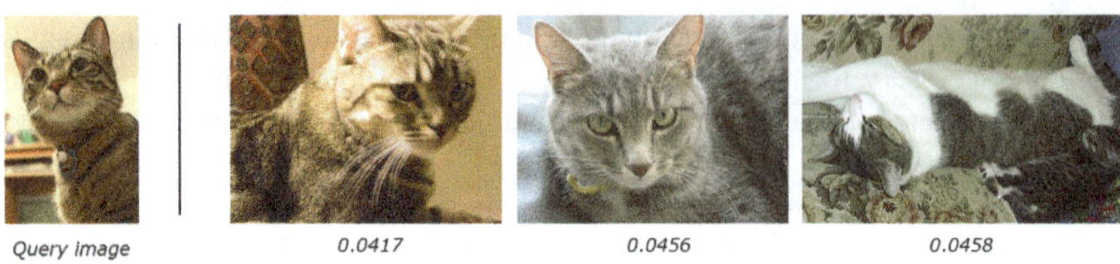

Fig. 5.6 Querying test performed for an image which is not included in the database. The distance to the query image is given for each returned image

Fig. 5.7 Querying test performed for an image which was included in the database. The distance to the query image is given for each returned image

5.2.3   Conclusions

We developed a system dedicated to image retrieval by providing an integrated environment for image analysis in a relational database management system environment. Nowadays RDBMS are used for collecting very large amount of data, thus it is crucial to integrate them with content-based visual querying methods. In the proposed system computations concerning visual similarity are encapsulated in the business logic of our system, users are only required to have knowledge about communication interfaces included in the proposed software. Applying database indexing methods affects positively speeding up the image retrieval.

Moreover, our system is integrated with .NET platform. The authors chose the WCF technology for providing the remote interaction with the system. MS SQL Sever allows to attach assemblies implemented in .NET to the database dedicated for image analysis. As a consequence, users can interact with the system locally by SQL commands, which execute remote procedures. It is an important advantage of the system. The system retrieves images in near real-time.

5.3   Database Indexing System Based on Boosting and Fuzzy Sets

In this section a novel method for fast classification of various images in large collections on the basis of their content, i.e. with a visual query-by-example problem in relational databases [10]. Namely, we describe a system based on Microsoft SQL Server which can classify a sample image or to return similar images to this image.

As aforementioned, users of the world most popular relational database management systems (RDBMS) such as Oracle, MS SQL Server and IBM DB2 Server are not encouraged to store such data directly in database files. An example of such an approach can be Microsoft SQL Server or Oracle where binary data is stored outside the RDBMS, and only the information about data location is stored in database tables. Those utilise a particular field type called FileStream or External tables (respectively), which integrate SQL Server database engine with the NTFS file system by storing large binary object (BLOB) data as files in its file system.

Since the release of SQL Server 2012, Microsoft added new functionality in the form of FileTable. This option is based on the filestream technology and also provides Windows users to access data file through a shared network resource. This allows access to the data file both from the MS SQL Server Application Programming Interface or from the NTFS file system. In operation, the data saved in the files can be accessed directly, at the same time relieving the database server.

In the proposed system, large image files are stored in special tables in SQL Server called FileTables as a FileStream field. MS SQL Server has a wide range of commands for working with large collections of text data (Full Text Search) and complex data - spatial coordinates.

The standard SQL language does not contain commands for handling multimedia. In addition, in the previous solutions, image files are stored directly in database tables. This has a significant impact on both low efficiency of the whole system and tasks related to the maintenance of a database, for example, time-consuming data backup which at the same time is very large in terms of volume. The proposed system also allows to inquiry the database similar SQL queries to

```
Select * from table_with_images
where fclass(image_field)='Cat'
```

Fclass is a user-defined function (UDF) that uses the method proposed in this section to return the set of images of a given class. Through the use of fuzzy logic [22, 23], visual image features and boosting meta learning we created a special index which can build a database query execution plan. In the next section, we describe the proposed database content-based image classification system.

5.3.1 Building Visual Index

Proposed RDBMS-based fast visual classification system is based on the idea presented in [10, 11, 27, 29, 32]. Here we use it to obtain the intervals of visual feature values that will be utilized in the database index. The detailed procedure is described in Sect. 3.1 and [11]. We create T^C weak classifiers t, $t = 1, ..., T^C$ in the form of fuzzy rules [22, 23]. After one run of the boosting procedure, we obtain a set of fuzzy rules (weak classifiers) with parameters determined for a given class. The procedure is then repeated for the rest of the visual classes, and afterwards, the system is ready to classify new images. When a new image is inserted into a database, a trigger executes and computes new visual features for this image in the form of the matrix

$$Q = \begin{bmatrix} q^1 \\ q^2 \\ \vdots \\ q^u \end{bmatrix} = \begin{bmatrix} q_1^1 & \cdots & q_N^1 \\ q_1^2 & \cdots & q_N^2 \\ & \vdots & \\ q_1^u & \cdots & q_N^u \end{bmatrix}. \tag{5.1}$$

On the basis of the computed features, we can classify the image by assigning the class label with the highest classifier answer

$$f(Q) = \arg\max_{c=1,\ldots,V} H^C(Q), \tag{5.2}$$

where

$$H^C(Q) = \sum_{t=1}^{T^C} \beta_t F_t(Q) \tag{5.3}$$

and $F_t(Q)$ is computed by checking the activation level of all the fuzzy rules and aggregating them

$$F_t(Q) = \mathop{S}_{j=1}^{u} (\mathop{T}_{n=1}^{N} G_{n,t}(q_n^j)), \tag{5.4}$$

where S and T are t-norm and t-conorm, respectively.

5.3.2 *Proposed Database Framework*

In this section, we describe the proposed database framework (see Fig. 5.8) that operates in a similar manner to Full-Text Search techniques.

Table `Stats` is used to store information about the progress of the algorithm in both the learning and retrieval mode. Images, both those that constitute the learning and the testing datasets are stored in the `ImagesFT` table of FileTable type. Fields `sift_keypoints, gaussParams` are of the previously created UDT type. Table `ImagesFT` is the FileTable type for storing image files.

To disburden database server the following, most computationally demanding operations for image analysis are moved to an operating system background service implemented as WCF:

1. KeyPoints Generation Process, responsible for generating keypoints by SIFT, SURF, etc.
2. Fuzzy Rule Generation Process, creates fuzzy rules for a given image class,
3. FuzzyClassifier Quality Rating, evaluates fuzzy rules.

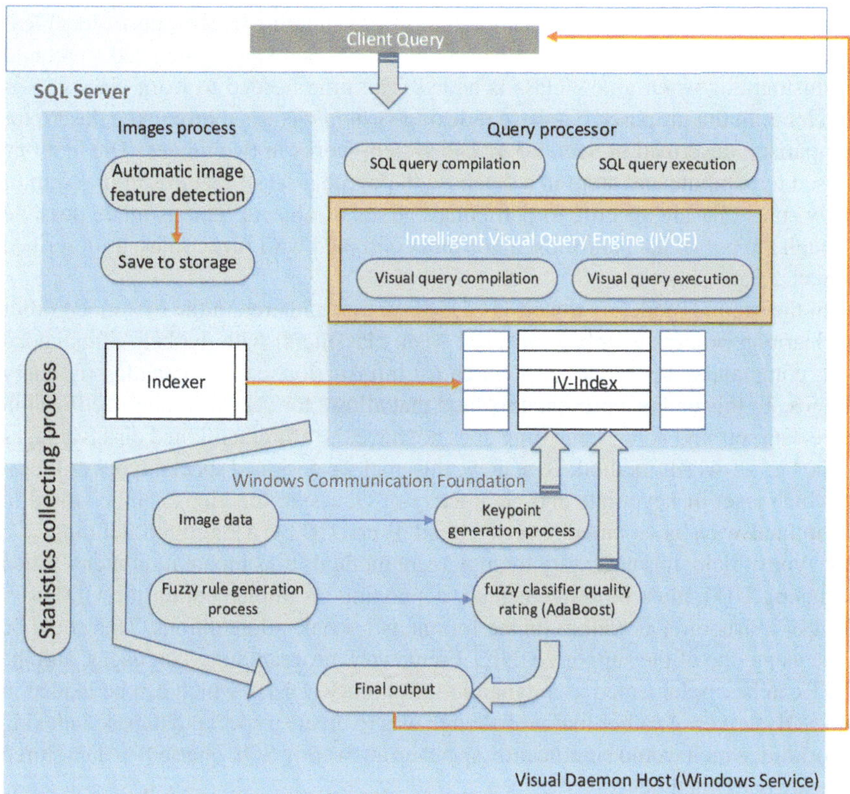

Fig. 5.8 Architecture of the proposed system (database and system processes with their mutual associations

This is only possible through the use of the FileTable technology, which makes possible to access image files through the API of the operating system without performing SQL queries. The indexing process itself takes place in the database environment. On the basis of fuzzy rules stored in the form of a dedicated type UDT, the Intelligent Visual Index is created, which is used to identify classes of images quickly or to search for images by the Intelligent Visual Query Engine module. The system starts indexing in the background immediately after inserting a visual object into the database. The primary information used to search for images in the presented system are SIFT keypoints, but the proposed method is universal enough to be able to use various visual features, for example, SURF or ORB. For a single image, we obtain a set of vectors of size 128. Generally, the task of finding two identical images comes down to comparing all vectors representing all images. Please note that in the case of large-scale sets of images, the problem is intractable in a detailed way and keypoint descriptors must be compared in an approximate way. One of the most popular methods currently used to solve the problem is the bag-of-features algorithm, where image retrieval or classification relies on comparing histograms of local features. Unfortunately, this method has some drawbacks. One is the need to rebuild the histograms when a new class is added. The time needed to learn classifiers is also long. In the proposed database indexing system we use approximate descriptor comparison described in Sect. 5.3.1. The system works in two modes. The first one is used to compute and store in a database the optimal set of features in the form of fuzzy rules that are specific to particular classes of objects. This is made possible through the use of the AdaBoost algorithm combined with fuzzy rules, as described in Sect. 5.3.1.

In the second mode, on the basis of information gathered in the first mode (after the learning process we obtain a set of weak classifiers) with the help of extended SQL commands, it is possible to search for information, e.g. we can classify query images. To this end, we present a method that allows for very fast image retrieval in large data sets and at the same time does not have the aforementioned disadvantages of the bag-of-words method. To achieve this goal we designed special types of fields, in which a set of keypoints and parameters of Gaussian functions can be stored in an optimal way, as so-called User-Defined Type (UDT). Along with defining of a new type of field, it is necessary to implement methods to compare its content. When designing UDT must be implemented, depending on implementing the UDT as a class or a structure, as well as on the format and serialisation options. This could be done using one of the supported .NET Framework programming languages, and the UDT can be implemented as a dynamic-link library (DLL), which can be loaded in MS SQL Server. Another major challenge was to create a special database indexing algorithm, which would significantly speed up answering SQL queries for data stored in a newly defined field.

After implementing the UDT, we added it as a library to the MS SQL Server environment. Then, we created the necessary tables for our system. Created UDT types contain all the information concerning the keypoint descriptors and Gaussian sets in a binary form. Thus, the database system cannot interpret the data stored there. For fast reference, in tables containing fields of the UDT type we added calculated fields to retrieve information from binary data, and then on their basis, the index is created. An example would be the `Gaussoids` table, where, among others, we extracted Gaussian sets parameters, the ranges in which the value of this function is greater than 0.5 from binary values stored in the `gaussParams` field of the type `GaussConfig`. Marking a computed column as PERSISTED allows an index to be created on a computed column that is deterministic, but not precise.

```
CREATE TABLE [dbo].[Gaussoids]
(
[config_id] INT IDENTITY (1, 1) NOT NULL,
[gaussParams] GaussConfig                 NOT NULL,
[C] AS gaussParams.C PERSISTED NOT NULL,
[ClassificatorId] AS gaussParams.ClassificatorId
            PERSISTED NOT NULL,
[PosGroupId] AS gaussParams.PosGroupId PERSIST-ED
        NOT NULL,
```

```
[InputNo] AS gaussParams.InputNo PERSISTED NOT NULL,
[RangeFrom] AS gaussParams.RangeFrom
            PERSISTED NOT NULL,
[RangeTo] AS gaussParams.RangeTo PERSISTED NOT NULL,
PRIMARY KEY CLUSTERED ([config_id] ASC)
);
```

At the beginning of the learning stage, we created a set of keypoints using the SIFT algorithm for every image (learning and testing set). Those vectors are stored as a `sift_keypoints` type fields in database tables. After this step, we created sets of rules for each image class. The result of the above procedure is the set of rules, which is then stored in the `Gaussoids` table. Please note, that by applying the Adaboost algorithm, each rule has been assigned a weight, i.e. a real number that indicates the quality of that rule in the classification process. This procedure allows us to identify the ranges in which a Gaussian function has a value greater than 0.5. Creating a database index on the fields `inputNo`, `RangeFrom` and `RangeTo` allows for fast determining which image feature values fall into ranges in which fuzzy sets which constitute the predecessor of the rule have values greater than 0.5. This situation is depicted in Fig. 5.9.

In the second mode we set up class labels for each of the image stored in the database, based on intervals obtained in the first mode.

When inserting an image into a FileTable-type table, which will be indicated by the user for indexing (in our system, we added ExtendedProperties called Key-PointsIndexed to such tables) there automatically starts the process of generating keypoint descriptors, which, as mentioned earlier, are stored in the form of UDT types dedicated to this table (Fig. 5.10). This action is imperceptible to the database user and is performed in a separate operating system process created in the WCF technology. Thus, despite the fact that the creation of a keypoint vector is very com-

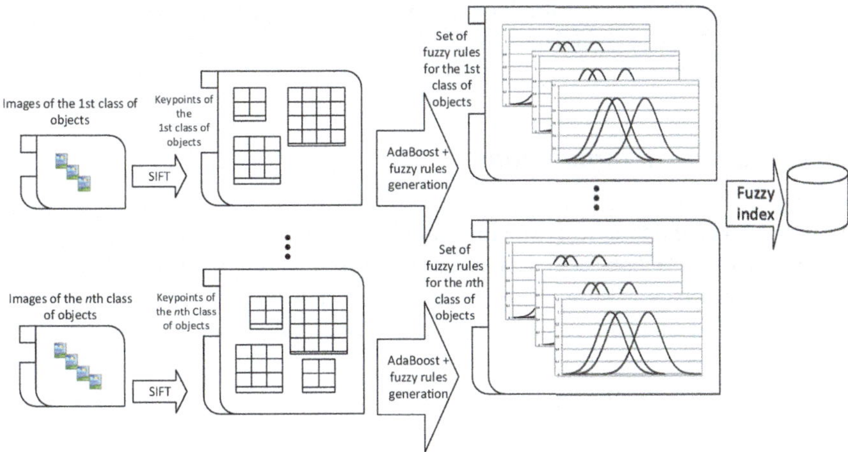

Fig. 5.9 Schema of index creating process

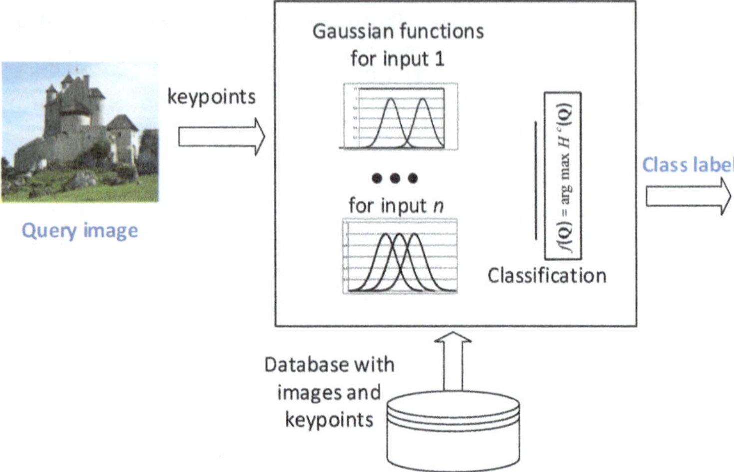

Fig. 5.10 Classification process for a new query image

putationally complex, it does not adversely affect the performance of the database itself. The classification process works in a similar manner. When a new image is inserted, the database trigger invokes a WCF function which checks membership of the image keypoint descriptors to individual rules.

According to [11], to compute the final answer of the system, i.e. image membership to a class, only rules which are activated at a minimum level 0.5 are taken into account. Thus, when using the minimum t-norm, only Gaussian sets in rule antecedents that are activated for the image keypoints to minimum 0.5 will have an impact on the image membership determination. Therefore, this information is stored in the database in the fields RangeFrom and RangeTo, with the database index set on these fields. This has a substantial impact on the search speed for specific Gaussian sets among millions of records.

5.3.3 Numerical Simulations

The proposed method was tested on four classes of visual objects taken from the PASCAL Visual Object Classes (VOC) dataset [5], namely: Bus, Cat, Dog and Train. The testing set consists of 15% of the images from the whole dataset. Before the learning procedure, we generated local keypoint vectors for all images from the Pascal VOC

Table 5.2 Experiments performed on images taken from the PASCAL Visual Object Classes (VOC) dataset for bag-of-features implementation with dictionary size 400 and various implementations of the proposed system

Implementation type	Testing time (s)	Learning time	Classification accuracy [%]
BoF on database	15.59	15m 30.00s	54.41
Desktop app.	14.44	10m 30.88s	54.41
RDBMS 1	9.41	10m 43.31s	52.94
RDBMS 2	8.93	10m 43.31s	51.40
RDBMS 3	2.50	10m 43.31s	57.35

dataset using the SIFT algorithm. All the experiments in this section were performed on a Hyper-V virtual machine with MS Windows Operating System (8 GB RAM, Intel Xeon X5650, 2.67 GHz). The testing set only contained images that had never been presented to the system during the learning process. We performed the experiments implementing the proposed content-based image classification algorithm as a desktop application written in C# language and as a database application, namely in Microsoft SQL Server. The goal was to show the advantages of using a database server for image content indexing. After training, we obtained a hundred rules for each visual class. Moreover, we compared the proposed method with the BOF algorithm implemented on the database server. The dictionary consisted of 400 visual words and was created outside the database. Then it was imported to the dedicated table. The classification accuracy was the same as in the case of RDBMS 1, but slower. Table 5.2 shows the execution times of the rule induction process and classification accuracy for desktop implementation of the proposed method and three versions of the database implementation (RDBMS 1 to 3). The methods named RDBMS 1 and RDBMS 2 used all the generated decision rules; however, RDBMS 2 used c_t to threshold the decision process. By the desktop application, we mean that simulations were made without database server means.

The best performance was achieved after merging similar decision rules into one rule with c_t being the sum of all merged c_t's (RDBMS 3). In this case, the system had fewer rules to check. We checked the rules against redundancy, and similar rules were merged into a new rule with c_t coefficient being the sum of the merged c_t. This operation allowed us to reduce computations for final classification substantially. In the RDBMS 1 method index is created only on the fields RangeFrom and RangeTo, whereas in RDBMS 2 and 3 we added the third field c_t.

We observe that by utilising database engine indexing connected with the proposed method, we can substantially speed up the retrieval process.

5.3.4 Conclusions

This section presents a new methodology for content-based image retrieval in relational databases based on a novel algorithm for generating fuzzy rules by boosting meta-learning. After learning, the parameters of fuzzy membership functions are used to create a database index for visual data. When new visual classes are introduced, the system generates a new, additional set of rules. Whereas in the case of other methods it would require a whole new dictionary generation and relearning of classifiers. The method uses the SIFT algorithm for visual feature computing, but it is possible to incorporate different features or different meta-learning algorithms. Image files are stored in the filesystem but are treated as database objects. This is convenient in terms of handling images with SQL queries and, at the same time, very fast when compared to the approaches presented in the literature.

Indispensable for the implementation of the presented algorithm is the database server to access image data not only through the database API, but also by the operating system API. In the presented case we used FileTable tables. In addition, the database server must have the ability to create extensions type UDT and UDF. It is not a serious limitation, because this condition is met in the most popular database systems. The solution, as shown in experimental results, does not have full accuracy. The accuracy is strongly dependent, as in most machine learning methods, on the quality of the images constituting training datasets and the parameters of the algorithm that generates the local image features. Performance of the whole solution can also be increased through the use of a SQL server cluster, where the process of generating the index in the form of rules can be parallelised and spread across several servers. Future directions would include the application of other visual features or methods of creating fuzzy rules and fuzzy sets.

5.4 Database Retrieval System Based on the CEDD Descriptor

In this section, we present a novel database architecture used to image indexing. The presented approach has several advantages over the existed ones:

- It is embedded into Database Management System (DBMS),
- Uses all the benefits of SQL and object-relational database management systems (ORDBMSs),
- It does not require any external program in order to manipulate data. A user of our index operate on T-SQL only, by using Data Modification Language (DML) by INSERT, UPDATE, and DELETE,

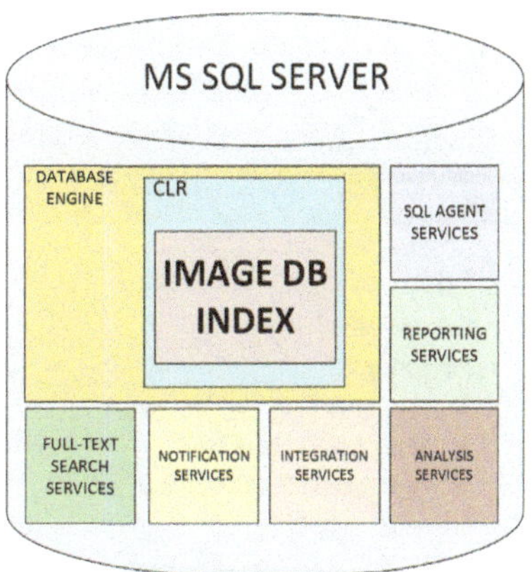

Fig. 5.11 The location of the presented image database index in Microsoft SQL Server

- Provides a new type for the database, which allows storing images along with the CEDD descriptor,
- It operates on binary data (vectors are converted to binary) thus, data processing is much faster as there is no JOIN clause used.

Our image database index is designed for Microsoft SQL Server, but it can be also ported to other platforms. A schema of the proposed system is presented in Fig. 5.11. It is embedded in the CLR (Common Language Runtime), which is a part of the database engine. After compilation, our solution is a .NET library, which is executed on CLR in the SQL Server. The complex calculations of the CEDD descriptor cannot be easily implemented in T-SQL thus, we decided to use the CLR C#, which allows implementing many complex mathematical transformations.

In our solution we use two tools:

- SQL C# User-Defined Types - it is a project for creating a user-defined types, which can be deployed on the SQL Server and used as the new type,
- SQL C# Function - it allows to create SQL Function in the form of C# code, it can also be deployed on the SQL Server and used as a regular T-SQL function. It should be noted that we use table-valued functions instead of scalar-valued functions.

At first we need to create a new user-defined type for storing binary data along with the CEDD descriptor. During this stage we encountered many issues which were resolved eventually. The most important ones are described below:

- The *Parse* method cannot take the *SqlBinary* type as a parameter, only *SqlString* is allowed. This method is used during INSERT clause. Thus, we resolve it by encoding binary to string and by passing it to the *Parse* method. In the body of the method we decode the string to binary and use it to obtain the descriptor,

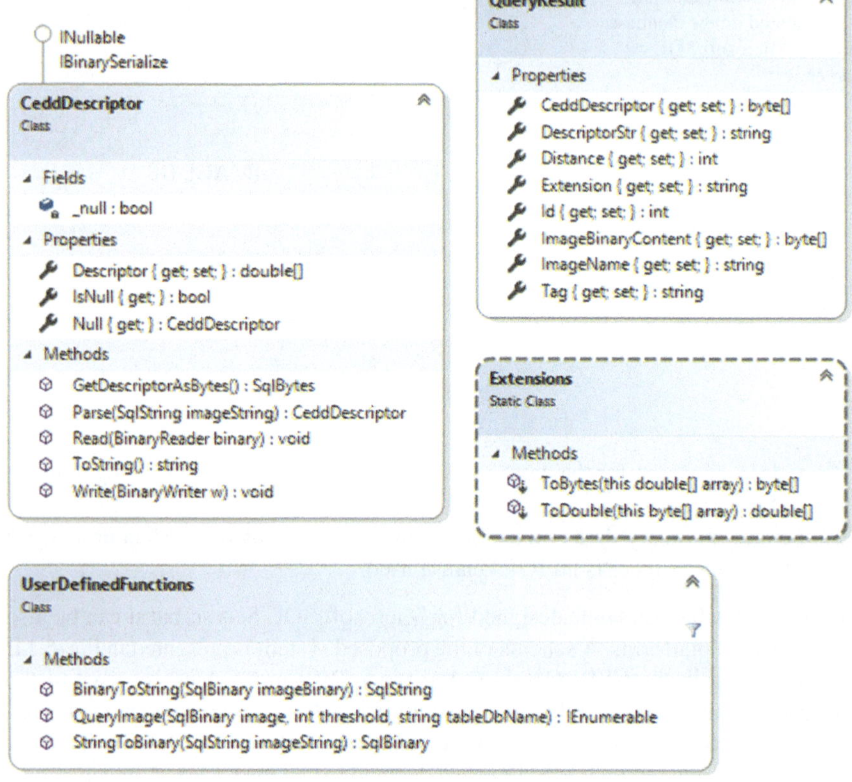

Fig. 5.12 Class diagram of the proposed database visual index

- Another interesting problem is registration of external libraries. By default the library $System.Drawing$ is not included. In order to include it we need to execute an SQL script.
- We cannot use reference types as fields or properties and we resolve this issue by implementing the $IBinarySerialize$ interface.

We designed one static class $Extensions$, and three classes: $CeddDescriptor$, $QueryResult$, $UserDefinedFunctions$ (Fig. 5.12). The $CeddDescriptor$ class implements two interfaces $INullable$ and $IBinarySerialize$. It also contains one field $\_null$ of type $bool$. The class also contains three properties and five methods. A $IsNull$ and $Null$ properties are required by user-defined types and they are mostly generated. The $Descriptor$ property allows to set or get the CEDD descriptor value in the form of a double array. A method $GetDescriptorAsBytes$ provides a descriptor in the form of a byte array. Another very important method is *Parse*. It is invoked automatically when the T-SQL *Cast* method is called (Listing 5.2). Due to the restrictions implemented in UDT, we cannot pass parameter of type *SqlBinary* as it must be *SqlString*. In order to resolve the nuisance we encode byte array to string by using the *BinaryToString* method from the *UserDefinedFunctions* class. In the body of the *Parse* method we decode the string to byte array, then we create a bitmap based on the previously obtained byte array. Next, the Cedd descriptor value is computed. Afterwards, the obtained descriptor is set as a property. The pseudo-code of this method is presented in Algorithm 4 The *Read* and *Write* method are implemented in order to use reference types as fields and properties. They are responsible for writing and reading to or from a stream of data. The last method (*ToString*) represents the *CeddDescriptor* as *string*. Each element of the descriptor is displayed as a string with a separator, this method allows to display the descriptor value by the SELECT clause.

INPUT: *EncodedString*
OUTPUT: *CeddDescriptor*
if *EncodedString = NULL* **then**
 | RETURN NULL;
end
ImageBinary := DecodeStringToBinary(EncodedString);
ImageBitmap := CreateBitmap(ImageBinary);
CeddDescriptor := CalculateCeddDescriptor(ImageBitmap);
SetAsPropertyDescriptor(CeddDescriptor)

Algorithm 4: Steps of the *Parse* method.

Another very important class is *UserDefinedFunctions*, it is composed of three methods. The *QueryImage* method performs the image query on the previously inserted images and retrieves the most similar images with respect to the *threshold* parameter. The method has three parameters: *image, threshold, tableDbName*. The first one is the query image in the form of a binary array, the second one determines the threshold distance between the image query and the retrieved images. The last parameter determines the table to execute the query on (it possible that many image tables exist in the system). The method takes the *image* parameter and calculates the *CeddDescriptor*. Then, it compares it with those existed in the database. In the next step the similar images are retrieved. The method allows filtering the retrieved images by the distance with the threshold. The two remaining methods *BinaryToString* and *StringToBinary* allow to encode and decode images as string or binary. The *QueryResult* class is used for presenting the query results to the user. All the properties are self-describing (see Fig. 5.12). The static *Extension* class contains two methods which extend double array and byte array, what allows to convert a byte array to a double array and vice versa.

5.4.1 Simulation Environment

The presented visual index was built and deployed on Microsoft SQL Server as a CLR DLL library written in C#. Thus, we needed to enable CLR integration on the server. Afterwards, we also needed to add *System.Drawing* and index assemblies as trusted. Then, we published the index and created a table with our new *CeddDescriptor* type. The table creation is presented on Listing 5.1. As can be seen, we created the *CeddDescriptor* column and other columns for the image meta-data (such as *ImageName*, *Extension* and *Tag*). The binary form of the image is stored in the *ImageBinaryContent* column.

Listing 5.1 Creating a table with the CeddDescriptor column.

```
CREATE TABLE CbirBow.dbo.CeddCorelImages
(
    Id int primary key identity(1,1),
    CeddDescriptor CeddDescriptor not null,
    ImageName varchar(max) not null,
    Extension varchar(10) not null,
    Tag varchar(max) not null,
    ImageBinaryContent varbinary(max) not null
);
```

Now we can insert data into the table what requires a binary data that will be loaded into a variable and passed as a parameter. This process is presented in Listing 5.2.

Listing 5.2 Inserting data to a table with the CeddDescriptor.

```
DECLARE @filedata AS varbinary(max);
SET @filedata = (SELECT *
FROM OPENROWSET(BULK N'{path_to_file}',
SINGLE_BLOB) as BinaryData)
INSERT INTO dbo.CeddCorelImages
(CeddDescriptor, ImageName, Extension, Tag,
        ImageBinaryContent)
VALUES (
CONVERT(CeddDescriptor, dbo.BinaryToString(@filedata)),
'644010.jpg', '.jpg', 'art_dino', @filedata);
```

Such prepared table can be used to insert images from any visual dataset, e.g. Corel, Pascal, ImageNet, etc. Afterwards, we can execute queries by the *QueryImage* method and retrieve images. For the experimental purposes, we used the PASCAL Visual Object Classes (VOC) dataset [5]. We split the image sets of each class into a training set of images for image description and indexing (90%) and evaluation, i.e. query images for testing (10%). In Table 5.3 we presented the retrieved factors of

Fig. 5.13   Example query results. The image with the border is the query image

multi-query. As can be seen, the results are satisfying which allows us to conclude that our method is effective and proves to be useful in CBIR techniques. For the purposes of the performance evaluation we used two well-known measures: *precision* and *recall* [16], see Sect. 3.2.

Figure 5.13 shows the visualization of experimental results from a single image query. As can be seen, most images were correctly retrieved. Some of them are improperly recognized because they have similar features such as shape or colour

Table 5.3 Simulation results (MultiQuery). Due to limited space, only a small part of the query results is presented

Image id	RI	AI	rai	iri	anr	Precision	Recall
598(pyramid)	50	47	33	17	14	66	70
599(pyramid)	51	47	31	20	16	61	66
600(revolver)	73	67	43	30	24	59	64
601(revolver)	72	67	41	31	26	57	61
602(revolver)	73	67	40	33	27	55	60
603(revolver)	73	67	42	31	25	58	63
604(revolver)	73	67	44	29	23	60	66
605(revolver)	71	67	40	31	27	56	60
606(revolver)	73	67	40	33	27	55	60
607(rhino)	53	49	39	14	10	74	80
608(rhino)	53	49	42	11	7	79	86
609(rhino)	53	49	42	11	7	79	86
610(rhino)	52	49	38	14	11	73	78
611(rhino)	52	49	39	13	10	75	80
612(rooster)	43	41	36	7	5	84	88
613(rooster)	43	41	33	10	8	77	80
614(rooster)	43	41	34	9	7	79	83
615(rooster)	44	41	35	9	6	80	85
616(saxophone)	36	33	26	10	7	72	79
617(saxophone)	36	33	26	10	7	72	79
618(saxophone)	35	33	26	9	7	74	79
619(schooner)	56	52	37	19	15	66	71
620(schooner)	56	52	37	19	15	66	71
621(schooner)	56	52	39	17	13	70	75
622(schooner)	55	52	37	18	15	67	71
623(schooner)	56	52	35	21	17	62	67
624(scissors)	35	33	22	13	11	63	67
625(scissors)	36	33	22	14	11	61	67
626(scissors)	36	33	20	16	13	56	61
627(scorpion)	75	69	59	16	10	79	86
628(scorpion)	73	69	57	16	12	78	83
629(scorpion)	73	69	58	15	11	79	84
630(scorpion)	73	69	59	14	10	81	86
631(scorpion)	74	69	55	19	14	74	80
632(scorpion)	75	69	56	19	13	75	81
633(scorpion)	74	69	53	21	16	72	77
634(sea-horse)	51	47	30	21	17	59	64
635(sea-horse)	51	47	30	21	17	59	64
636(sea-horse)	50	47	29	21	18	58	62

(continued)

Table 5.1 (continued)

Image id	RI	AI	rai	iri	anr	Precision	Recall
637(sea-horse)	50	47	32	18	15	64	68
638(sea-horse)	49	47	30	19	17	61	64
639(snoopy)	31	29	24	7	5	77	83
640(snoopy)	31	29	22	9	7	71	76
641(snoopy)	31	29	22	9	7	71	76
642(soccer-ball)	56	53	43	13	10	77	81
643(soccer-ball)	57	53	44	13	9	77	83
644(soccer-ball)	56	53	42	14	11	75	79
645(soccer-ball)	57	53	46	11	7	81	87
647(stapler)	40	37	32	8	5	80	86
Average						71	76

background. The image with the red border is the query image. The *Average Precision* value for the entire dataset equals 71 and for *AverageRecall* 76.

5.4.2 Conclusions

The presented system is a novel architecture of a database index for content-based image retrieval. We used Microsoft SQL Server as the core of our architecture. The approach has several advantages: it is embedded into RDBMS, it benefits from the SQL commands, thus it does not require external applications to manipulate data, and finally, it provides a new type for DBMSs. The proposed architecture can be ported to other DBMSs (or ORDBMSs). It is dedicated to being used as a database with CBIR feature. The performed experiments proved the effectiveness of our architecture. The proposed solution uses the CEDD descriptor but it is open to modifications and can be relatively easily extended to other types of visual feature descriptors.

5.5 Summary and Discussion

This chapter presented several implementations of content-based image retrieval and classification systems in relational database management systems. A process associated with retrieving images in the databases is query formulation (similar to the SELECT statement in the SQL language). All the presented systems are operated on the query-by-image principle. Survey [15] mentions three visual query levels:

1. Level 1: Retrieval based on primary features like colour, texture and shape. A typical query is "search for a similar image".

2. Level 2: Retrieval of a certain object which is identified by extracted features, e.g. "search for a flower image".
3. Level 3: Retrieval of abstract attributes, including a vast number of determiners about the presented objects and scenes. Here, it is possible to find names of events and emotions. An example query is: "search for satisfied people".

The first method in this chapter presented a fast content-based image classification algorithm implemented in a relational database management system using the bag of features approach, Support Vector Machine classifiers and special Microsoft SQL Server features. Moreover, a fuzzy index was designed to search for similar images to the query image in large sets of visual records. The described framework allows automatic searching and retrieving images on the base of their content using the SQL language. The SQL responses are nearly real-time with even relatively large image datasets.

Next, two systems based on the bag of words approach for retrieving and classifying images as integrated environments for image analysis in a relational database management system environment. In the proposed systems computations concerning visual similarity are encapsulated in the business logic of our system, users are only required to have knowledge about communication interfaces included in the proposed software. Users can interact with the systems locally by SQL commands, which execute remote procedures. It is an important advantage of the system. Image files are stored in the filesystem but are treated as database objects. This is convenient in terms of handling images with SQL queries and, at the same time, very fast when compared to the approaches presented in the literature. The systems retrieve images in near real-time.

Finally a novel architecture of a database index for content-based image retrieval with Microsoft SQL Server based on the CEDD descriptor. The proposed architecture can be ported to other DBMSs (or ORDBMSs). It is dedicated to being used as a database with CBIR feature. The proposed solution uses the CEDD descriptor; however, it is open to modifications and can be relatively easily extended to other types of visual feature descriptors. The system can be extended to use different visual features or to have a more flexible SQL querying command set. The performed experiments proved the effectiveness of the architectures. The presented systems can be a base for developing more sophisticated querying by incorporating natural language processing algorithms.

References

1. Araujo, M.R., Traina, A.J., Traina C., Jr.: Extending SQL to support image content-based retrieval. In: ISDB, pp. 19–24 (2002)
2. Bradski, G.: The opencv library. Dr. Dobbs J. **25**(11), 120–126 (2000)
3. Chaudhuri, S., Narasayya, V.R.: An efficient, cost-driven index selection tool for microsoft SQL server. VLDB **97**, 146–155 (1997)

4. Dubois, D., Prade, H., Sedes, F.: Fuzzy logic techniques in multimedia database querying: a preliminary investigation of the potentials. IEEE Trans. Knowl. Data Eng. **13**(3), 383–392 (2001). https://doi.org/10.1109/69.929896
5. Everingham, M., Van Gool, L., Williams, C.K.I., Winn, J., Zisserman, A.: The pascal visual object classes (VOC) challenge. Int. J. Comput. Vis. **88**(2), 303–338 (2010)
6. Fielding, R.T.: Architectural styles and the design of network-based software architectures. Ph.D. thesis, University of California, Irvine (2000)
7. Grauman, K., Darrell, T.: Efficient image matching with distributions of local invariant features. In: IEEE Computer Society Conference on Computer Vision and Pattern Recognition, 2005. CVPR 2005, vol. 2, pp. 627–634 (2005). https://doi.org/10.1109/CVPR.2005.138
8. Kacprzyk, J., Zadrozny, S.: Fuzzy queries in microsoft access v. 2. In: Proceedings of the FUZZ-IEEE/IFES'95 Workshop on Fuzzy Database Systems and Information Retrieval (1995)
9. Kacprzyk, J., Zadrozny, S.: On combining intelligent querying and data mining using fuzzy logic concepts. In: Recent Issues on Fuzzy Databases, pp. 67–81. Springer (2000)
10. Korytkowski, M.: Novel visual information indexing in relational databases. Integr. Comput. Aided Eng. **24**(2), 119–128 (2017)
11. Korytkowski, M., Rutkowski, L., Scherer, R.: Fast image classification by boosting fuzzy classifiers. Inf. Sci. **327**, 175–182 (2016). https://doi.org/10.1016/j.ins.2015.08.030. http://www.sciencedirect.com/science/article/pii/S0020025515006180
12. Korytkowski, M., Scherer, R., Staszewski, P., Woldan, P.: Bag-of-features image indexing and classification in microsoft sql server relational database. In: 2015 IEEE 2nd International Conference on Cybernetics (CYBCONF), pp. 478–482 (2015). https://doi.org/10.1109/CYBConf.2015.7175981
13. Larson, P.Å., Clinciu, C., Hanson, E.N., Oks, A., Price, S.L., Rangarajan, S., Surna, A., Zhou, Q.: SQL server column store indexes. In: Proceedings of the 2011 ACM SIGMOD International Conference on Management of Data, pp. 1177–1184. ACM (2011)
14. Liu, J.: Image retrieval based on bag-of-words model (2013). arXiv:1304.5168
15. Liu, Y., Zhang, D., Lu, G., Ma, W.Y.: A survey of content-based image retrieval with high-level semantics. Pattern Recognit. **40**(1), 262–282 (2007)
16. Meskaldji, K., Boucherkha, S., Chikhi, S.: Color quantization and its impact on color histogram based image retrieval accuracy. In: First International Conference on Networked Digital Technologies, 2009. NDT'09, pp. 515–517 (2009). https://doi.org/10.1109/NDT.2009.5272135
17. Müller, H., Geissbuhler, A., Marchand-Maillet, S.: Extensions to the multimedia retrieval markup language–a communication protocol for content–based image retrieval. In: European Conference on Content-based Multimedia Indexing (CBMI03). Citeseer (2003)
18. Ogle, V.E., Stonebraker, M.: Chabot: retrieval from a relational database of images. Computer **9**, 40–48 (1995)
19. Pein, R.P., Lu, J., Renz, W.: An extensible query language for content based image retrieval based on Lucene. In: 8th IEEE International Conference on Computer and Information Technology, 2008. CIT 2008, pp. 179–184. IEEE (2008)
20. Philbin, J., Chum, O., Isard, M., Sivic, J., Zisserman, A.: Object retrieval with large vocabularies and fast spatial matching. In: IEEE Conference on Computer Vision and Pattern Recognition, 2007. CVPR'07, pp. 1–8 (2007)
21. Rivest, R.: The MD5 Message-Digest Algorithm. RFC Editor, United States (1992)
22. Rutkowski, L.: Computational Intelligence Methods and Techniques. Springer, Berlin (2008)
23. Scherer, R.: Multiple Fuzzy Classification Systems. Springer (2012)
24. Sivic, J., Zisserman, A.: Video google: a text retrieval approach to object matching in videos. In: Proceedings of the Ninth IEEE International Conference on Computer Vision, 2003, vol. 2, pp. 1470–1477 (2003)
25. Srinivasan, J., De Fazio, S., Nori, A., Das, S., Freiwald, C., Banerjee, J.: Index with entries that store the key of a row and all non-key values of the row (2000). US Patent 6,128,610
26. Staszewski, P., Woldan, P., Korytkowski, M., Scherer, R., Wang, L.: Artificial Intelligence and Soft Computing: 15th International Conference, ICAISC 2016, Zakopane, Poland, June 12–16, (2016) (Proceedings, Part II, chap. Query-by-Example Image Retrieval in Microsoft SQL Server, pp. 746–754. Springer International Publishing, Cham, 2016)

27. Tieu, K., Viola, P.: Boosting image retrieval. Int. J. Comput. Vision **56**(1–2), 17–36 (2004)
28. Vagač, M., Meličerčík, M.: Improving image processing performance using database user-defined functions. In: International Conference on Artificial Intelligence and Soft Computing, pp. 789–799. Springer (2015)
29. Viola, P., Jones, M.: Rapid object detection using a boosted cascade of simple features. In: Proceedings of the 2001 IEEE Computer Society Conference on Computer Vision and Pattern Recognition, 2001. CVPR 2001, vol. 1, pp. I–511–I–518 (2001)
30. Voloshynovskiy, S., Diephuis, M., Kostadinov, D., Farhadzadeh, F., Holotyak, T.: On accuracy, robustness, and security of bag-of-word search systems. In: IS&T/SPIE Electronic Imaging, pp. 902, 807–902, 807. International Society for Optics and Photonics (2014)
31. Zhang, J., Marszalek, M., Lazebnik, S., Schmid, C.: Local features and kernels for classification of texture and object categories: a comprehensive study. In: Conference on Computer Vision and Pattern Recognition Workshop, 2006. CVPRW'06, pp. 13–13 (2006). https://doi.org/10. 1109/CVPRW.2006.121
32. Zhang, W., Yu, B., Zelinsky, G.J., Samaras, D.: Object class recognition using multiple layer boosting with heterogeneous features. In: 2005 IEEE Computer Society Conference on Computer Vision and Pattern Recognition (CVPR'05), vol. 2, pp. 323–330 (2005). https://doi.org/ 10.1109/CVPR.2005.251

Chapter 6
Concluding Remarks and Perspectives in Computer Vision

The previous chapters covered some topics relating to computer vision: how global and local features are generated, how to fast index them and how to implement content-based retrieval algorithms in relational database management systems. Chapter 1 is an introduction to the book subject. Chapter 2 presents several methods for image feature detection and description, starting from image interest points, through edge and blob detection, image segmentation till global features. Chapter 3 concerns feature comparison and indexing for efficient image retrieval and classification. Chapter 4 presents novel methods for feature description and Chap. 5 consists of a set of relational database implementation. Computer vision is not a mature discipline and is continually developing and evolving. Therefore, it is not possible to cover all the directions and solve all challenges within the scope of one book. Currently, it is hard to rival human vision in a general sense as it is our most powerful sense. Deep learning and hardware rapid development gradually change this situation. In 2015 neural networks defeated humans in the ImageNet Large Scale Visual Recognition Challenge. Computer vision starts to shift from relying on hand-made features to learned features. This can constitute a direction in the future research, namely, using trained features in the methods described in Chaps. 3 and 5, would possibly improve the accuracy. Moreover, the robustness in terms of immunity to noise, occlusions, distortion, shadows etc. can also be improved. Computer vision benefits heavily from the development of computer hardware as many algorithms are NP-complete. Since Moore's law (and other types of the hardware development) will most likely still be valid, vision system will be more and more sophisticated.

© Springer Nature Switzerland AG 2020

R. Scherer, *Computer Vision Methods for Fast Image Classification and Retrieval*, Studies in Computational Intelligence 821,
https://doi.org/10.1007/978-3-030-12195-2_6

The manufacturer's authorised representative in the EU is Springer
Nature Customer Service Centre GmbH, Europaplatz 3, 69115 Heidelberg,
Germany. If you have any concerns regarding our products, please
contact ProductSafety@springernature.com

Printed and bound by CPI Group (UK) Ltd, Croydon, CR0 4YY
28/11/2025
02007724-0006